The Political Economy of Trust

Institutions, Interests, and Inter-Firm Cooperation in Italy and Germany

Trust and cooperation are at the heart of the two most important approaches to comparative politics – rational choice and political culture. Yet we know little about the relationship of trust to political institutions. This book sets out a rationalist theory of how informal institutions can affect trust without reducing it to fully determined expectations. It then shows how this theory can be applied to comparative political economy, in particular to explaining inter-firm cooperation in industrial districts, geographical areas of intense small firm collaboration. The book compares trust and cooperation in two prominent districts in the literature, Emilia-Romagna, Italy, and Baden-Württemberg, Germany. It also sets out and applies a theory of how national informal institutions may change as a result of changes in global markets, and it shows how similar mechanisms may explain persistent distrust among Sicilian *mafiosi.*

Henry Farrell has been an assistant professor of political science and international relations since 2004 at the George Washington University, where he is affiliated with the Center for International Science and Technology Policy. He was an assistant professor at the University of Toronto from 2002 to 2004 and Senior Research Fellow at the Max-Planck Project Group on the Law of Common Goods in Bonn, Germany, from 2000 to 2002. He has authored or co-authored 20 peer-reviewed articles in scholarly journals, including *International Organization* and *Comparative Political Studies*, and ten chapters for edited volumes. His nonacademic publications include articles for *Foreign Policy*, *Financial Times*, *Boston Review*, *Chronicle of Higher Education*, *American Prospect*, and *The Nation*, as well as a policy paper for the German Marshall Fund.

Other Books in the Series:

David Austen-Smith, Jeffrey A. Frieden, Miriam A. Golden, Karl Ove Moene, and Adam Przeworski, eds., *Selected Works of Michael Wallerstein: The Political Economy of Inequality, Unions, and Social Democracy*
Lisa Baldez, *Why Women Protest: Women's Movements in Chile*
Stefano Bartolini, *The Political Mobilization of the European Left, 1860–1980: The Class Cleavage*
Robert Bates, *When Things Fell Apart: State Failure in Late-Century Africa*
Mark Beissinger, *Nationalist Mobilization and the Collapse of the Soviet State*

Continued after the Index

The Political Economy of Trust

Institutions, Interests, and Inter-Firm Cooperation in Italy and Germany

HENRY FARRELL
George Washington University

CAMBRIDGE
UNIVERSITY PRESS

CAMBRIDGE UNIVERSITY PRESS
Cambridge, New York, Melbourne, Madrid, Cape Town, Singapore, São Paulo, Delhi

Cambridge University Press
32 Avenue of the Americas, New York, NY 10013–2473, USA

www.cambridge.org
Information on this title: www.cambridge.org/9780521886499

First published 2009

Printed in the United States of America

A catalog record for this publication is available from the British Library.

Library of Congress Cataloging in Publication data
Farrell, Henry, 1970–
The political economy of trust : institutions, interests, and inter-firm cooperation in Italy and Germany / Henry Farrell.
p. cm. – (Cambridge studies in comparative politics)
Includes bibliographical references and index.
ISBN 978-0-521-88649-9 (hardback)
1. Business networks – Italy – Case studies. 2. Business networks – Germany – Case studies. 3. Strategic alliances (Business) – Italy – Case studies. 4. Strategic alliances (Business) – Germany – Case studies. 5. Rational choice theory – Case studies. 6. Packaging machinery industry – Italy – Emilia-Romagna. 7. Machine-tool industry – Germany – Baden-Württemberg. 8. Mafia – Italy – Sicily. I. Title. II. Series.
HD69.S8F26 2009
338.8′70943–dc22 2008044094

ISBN 978-0-521-88649-9 hardback

Contents

Acknowledgments

What is the relationship between institutions and trust? This question, which has fascinated me for many years, spurred both my dissertation and this (very different and, I hope, much improved) book. My fascination began when my academic mentor, Sam Barnes, suggested that I read Robert Putnam's 1993 book, *Making Democracy Work*. Like many others, I was captivated by the breadth and ambition of the book, which sought to reconcile arguments from political culture, the new institutional economics, Renaissance history, and many other sources to come up with a grand theory of the basic underpinnings of politics. I was also dissatisfied with Putnam's account of trust and cooperation in Italy in ways that I at first had difficulty describing, even to myself, and I started to pull, in a not very organized way, on loose threads to see what might give.

As I began to define my question better, I became aware of the extensive literature in game theory and rational choice on institutions and cooperation, in large part as a result of the tutelage of Jim Johnson. Yet I found this literature unsatisfactory too, especially when it tried to explain the more diffuse kinds of trust and cooperation that Putnam and others had written about. This book represents my efforts to come up with a better answer than either approach and, more generally, to try to use the rational choice theory of institutions as a way of understanding how individuals could come to trust and cooperate with each other, not only in the well-specified ways that game theoretic accounts describe but also in fuzzier situations, which are not clearly defined.

I owe intellectual and personal debts to many people. I also, to my very happy surprise, found that my intellectual conversations and research collaborations concerning this topic became the seedbed for many friendships. This is something that seems obvious now, but it was not at all obvious to me when I began as a graduate student to research this topic and write about it. When you enter into a community of argument, you find that it is a community in more than one way.

First, and most importantly, I am enormously grateful to my dissertation committee. I owe a great intellectual debt to my graduate school mentor and dissertation adviser, Sam Barnes. Not only did he provide me with a fine academic grounding in the understanding of trust and in Italian politics, but he also gave me free rein to build out from these foundations in unorthodox directions. Colin Crouch provided me with more specific guidance in comparative political economy and with a broader institutional framework in which I could pursue my particular interests. Margaret Levi gave me a much needed grounding in the literature on both trust and institutions. Mark Warren provided me with comments that helped me to see both the broader implications and more specific limitations of my argument.

In addition, I owe specific debts to Jack Knight and Jim Johnson. Many arguments presented in this book emerged from my running debate with Jack on trust and institutional theory over the last few years – we have written together on this and related topics, and we plan to write more in the future. As I have mentioned, Jim Johnson's ICPSR course on rational choice provided me with many of the intellectual tools that I have used (or abused!) to write this book.

My original research for this book was carried out under the auspices of a larger project on the organization of local economies in Europe directed by Colin Crouch, Patrick le Galès, Carlo Trigilia, and Helmut Voelzkow. I am grateful to them for having given me the opportunity to piggyback my research interests on top of this larger project. I am also grateful to the two people with whom I worked most closely on the project, Ulrich Glassmann and Ann-Louise Holten. Ulrich carried out the lion's share of the research in Germany, while Ann-Louise conducted interviews with Italian firms. I am grateful to them not only for their work with me on this topic but also for the many conversations that have informed the ideas and claims set out in this book. Luigi Burroni also provided help and support throughout the process of writing this book. I am eternally grateful to him. Chapter 4 of this book was originally published as Henry Farrell, 2005, "Trust and Political Economy: Comparing the Effects of Institutions on Inter-Firm Cooperation," *Comparative Political Studies* 38(5):459–483. A much earlier version of Chapter 5 was published as Henry Farrell and Jack Knight, 2003, "Trust, Institutions and Institutional Change: Industrial Districts and the Social Capital Hypothesis," *Politics and Society* 31(4): 537–566. The newer version interprets the evidence differently: any mistakes in the earlier version are mine, not my co-author's.

Russell Hardin's ideas are omnipresent in this book; my claims owe much to his accounts (developed in his solo-authored work and his work in collaboration with Karen Cook and Margaret Levi) of the encapsulated interest account of trust. He has been extraordinarily intellectually generous. Diego Gambetta deserves considerable thanks both for his generosity in making his database of Mafia-related information generally available and for some very helpful e-mail conversations about the Mafia and trust. Both Gary Herrigel and Jonathan Zeitlin provided me with good advice and tough, useful criticisms at early stages of this project. While both disagreed (and still disagree) vigorously with many of my claims, their intellectual generosity helped sharpen my arguments. Sven Steinmo's conversation with me in a Brussels cafeteria also helped shape this project by forcing me to clarify my claims.

I am also grateful to Stefano Bartolini, Marco Bellandi, Nicola Bellini, the late Hans-Joachim Braczyk, Randy Calvert, Jim Caporaso, Karl Cerny, Roger Chickering, Karen Cook, Tyler Cowen, Gabi dei Ottati, Bruce Douglass, Mario Drago, Jean Ensminger, Neil Fligstein, Tom Garvin, Avner Greif, Lois Harder, Kieran Healy, Adrienne Heritièr, Joe Jupille, John Kenny, Peter Klein, Gerhard Krauss, Dawn Lyon, Rosarie McCarthy, Moore McDowell, Gary Miller, Kimberly Morgan, Maria Murray, the late Mancur Olson, Lin Ostrom, Vincent Ostrom, Gabi dei Ottati, Gianfranco Poggi, Margherita Russo, Melissa Schwartzberg, Richard Sinnott, Sven Steinmo, Alec Stone Sweet, Carlo Trigilia, Josh Whitford, and Jonathan Zeitlin. All of these people provided helpful readings, criticisms, or feedback at different stages of the argument. Melissa Schwartzberg also provided much moral support as we both tried to finish our books for Cambridge. Two anonymous reviewers provided excellent, tough comments that have resulted in considerable improvements to the manuscript.

I am grateful to the European University Institute in Florence, for bringing me in as a researcher and for treating me as one of their own during the larger period of my dissertation research. I am also grateful to the Max-Planck Institute for the Study of Society for hosting me for a month, and to the Akademie für Technikfolgenabschätzung for providing me with facilities and a happy work environment during my time in Stuttgart. Both the Max-Planck Project Group on the Law of Common Goods in Bonn and the Department of Political Science of the University of Toronto provided congenial work environments as I began to turn my research into a book. I owe a great debt of gratitude to all my former colleagues. My current intellectual home, the George Washington University, has provided me the resources to finish the book and membership in a thriving intellectual community.

The people at Cambridge University Press have provided me with considerable assistance and support as I wrote this book. I owe an especial debt of gratitude to my editor, Lew Bateman, both for patience and for intellectual support and advice throughout. Regina Paleski did a wonderful job as production editor in seeing this book reach press. Emily Spangler provided considerable help in coordinating the book cover and other matters. Finally, Scott Barker provided excellent copyediting, not only correcting for stylistic tics and English language errors but also, to my surprise and delight, spotting and correcting a grammatical mistake in Italian, too.

Last, but most important, family: I cannot express my gratitude to my wife, Nicole Erb, and our sons, Jack and Kieran. Nicole came into my life just after I had finished my dissertation; this book has been in gestation ever since. She has been extraordinarily patient and helpful in giving me time and space to complete it. Jack and Kieran have contributed in a more inadvertent fashion, through the dates on which they arrived in this world; Jack's imminent arrival pushed me to complete the first major draft, and Kieran's arrival (some 10 days before I write these words) pushed me to complete this one. I owe particular thanks to my parents and siblings, my late grandmother Eilis McDowell, and my aunt Niamh. Finally, this book is dedicated to my late aunt and godmother, Bairbre O'Dwyer, in fulfilment of a promise made long ago. She was never happy with half measures and asked, when I made her this promise, whether I was going to write an *interesting* book. I can only hope that this book would have been interesting enough to satisfy her formidable intellect, and I trust that her warmth and kindness would have led her to overlook the many flaws I am sure still remain.

I

Introduction

1.1 A POLITICAL ECONOMY OF TRUST

What are the sources of trust and cooperation in political and economic life? This question lies at the heart of key debates within political science. However, it remains unresolved, in large part because of serious weaknesses in our understanding of the relationship between social structure, trust, and cooperation.

These weaknesses are not the result of neglect by political scientists. The two dominant approaches to the explanation of comparative politics, political culture and rational choice, both place trust and cooperation at the core of their research agenda.[1] Yet there is something unsatisfying about the arguments that they offer. Scholars of political culture have almost uniformly concerned themselves with macrolevel arguments based on aggregate survey data. Their arguments have scanty microfoundations, a rather serious problem if one wishes to explain how individuals may come to trust each other and thus cooperate. Rational choice scholars have constructed models with clearly specified microfoundations, but they have had great difficulty to date

[1] I do not include historical institutionalism as a separate account of political behavior, as different variants borrow their microfoundations from both rational choice and culturalist accounts of politics. See Hall and Taylor 1996. More recent varieties of historical institutionalism build upon results from path dependence theory (Pierson 2000; 2004), but they do not have unique microfoundations. However, see Chapter 2 for a more detailed discussion of historical institutionalist claims.

in building mid-level theories that can capture the consequences of institutions and other such broader phenomena for trust. All too often, rational choice scholars implicitly or explicitly equate trust with simple institution-induced expectations,[2] a move which nonrational choice scholars (correctly) view as reductionist.

In this book, I try to set out the beginnings of an alternative approach to the understanding of the "political economy" of trust; that is, of the relationship between institutions, trust, and cooperation in economic interactions. My account builds upon the basic foundations of rational choice theory. The above-stated caveats aside, I argue that rational choice provides excellent building blocks for a coherent and persuasive theory of what motivates actors to trust each other in economic situations, and thus to cooperate. On the one hand, scholars such as Douglass North (1990) and Jack Knight (1992) provide us with the means to analyze institutions as sets of rules that can shape individual behavior. On the other hand, recent work by Russell Hardin (2002) and others, most particularly Margaret Levi (1998) and Cook, Hardin, and Levi (2005), on the "encapsulated interest" account of trust and related ideas, allows us to see how individuals may draw inferences about the trustworthiness of others from the interests that those others have in behaving trustworthily. In other words, rational choice has the two basic components for the kind of theory that I seek to construct: it provides arguments about (1) the origins and nature of institutions and (2) the bases of trust and social cooperation.

Even so, it is difficult to draw these two literatures within rational choice together into a coherent whole without falling into the trap of reductionism. How may institutions affect the possibility for relations based on trust without, at the same time, entirely determining them? If everything is explained by institutional incentives, then it is difficult to talk about trust in any meaningful way. We have a powerful (and, I believe, valid) intuition that trust encompasses something rather richer than the mechanical sets of incentives and relationships that formal models depict. In two important contributions, David Kreps (1990a) and Gary Miller (1992) provide the beginnings of a more fruitful line of inquiry, which builds on game theory to understand trust in a less mechanistic fashion. In this book, I seek to build on

[2] For an especially trenchant statement of this claim, see Jackman and Miller 2004.

their insights, to integrate them with recent advances in institutional theory, and then to apply them to the comparative political economy of inter-firm relations.

The relationship between institutions, trust, and cooperation is an important one for many avenues of inquiry in the social sciences. I argue that it has a particular relevance for comparative political economy. A recent landmark essay by Peter Hall and David Soskice (2001) endeavours to set out the beginnings of a general comparative approach to political economy, building on both historic institutionalism and rational choice theory. I hope to contribute to the "varieties of capitalism" literature that Hall and Soskice wish to systematize, albeit with a different understanding of the forces that guide institutional change and the likely consequences of institutions for individual behavior.

In order to develop this contribution properly, I make use of two rich bodies of empirical evidence. First, I turn to two case studies of trust and cooperation within the political economy: the packaging machinery industry in Emilia-Romagna in Italy, and the machine-tool industry in Baden-Württemberg in Germany. These cases are important examples of "industrial districts," areas of small firm concentration, which, according to the existing scholarly literature, are characterized by extraordinarily high levels of cooperation among firms. In the classic industrial districts, intensive inter-firm cooperation appears to serve as a substitute for hierarchical production within the firm, a pattern of production that is both extraordinarily rare in advanced industrial economies (Miller 1992) and difficult to explain using standard approaches to the relationship between hierarchy and market (Williamson 1975; 1985). Furthermore, some scholars have claimed that this cooperation involves a kind of trust that cannot be explained using rational choice theory (Brusco 1990; Sabel 1993). In short, industrial districts present important puzzles for both political science and rational choice theory. Emilia-Romagna and Baden-Württemberg are among the "classic cases" of the industrial district phenomenon (Brusco 1982; 1990; Burroni and Trigilia 2001), and thus provide especially important test cases, while also providing a high degree of variation on institutions. Thus, they allow one to study the effects of institutions on trust and cooperation among actors with a reasonable degree of clarity and precision.

Second, I examine an empirical context characterized by widespread distrust and opportunism – relations within the Sicilian Mafia. For obvious reasons, this case is not directly comparable to mechanical engineering – the problems of trust are rather starker for a mafioso seeking to determine whether his apparent friend wishes to lure him to a secluded spot so as to strangle him, than for a producer of packaging machines trying to determine which subcontractor is trustworthy. Even so, it provides a parallel demonstration of how the mechanisms of expectation formation and information transmission that serve to underpin trust in one set of circumstances can help lead to pervasive distrust in another.

I use empirical evidence drawn from these cases to assay my broader theoretical claims. These case studies, even if they exemplify quite important phenomena, provide only an imperfect means of testing generalizable arguments. Yet there are some very real intellectual advantages to proceeding in this fashion. Robert Bates, Avner Greif, Margaret Levi, Jean-Laurent Rosenthal, and Barry Weingast (1998a) recommend the use of "analytic narratives," a combination of abstract theoretical argument, and detailed exegesis of empirical facts, to disentangle the causal relationships in complex strategic situations. Analytic narratives, as described by these scholars, are problem driven rather than theory driven; they seek to account for particular outcomes rather than to establish general causal relationships. Yet, as Peter Hall (2003) notes, studies of this sort may also serve wider theoretical goals beyond the specific explication of the particular cases under examination.

Indeed, such studies may be quite essential to the mapping of certain kinds of causal relationships. Hall suggests that the methodologies used by comparative social scientists have increasingly fallen out of step with the underlying ontologies that they assume. Cross-national multiple *n* multivariate analysis rests on assumptions about the underlying data such as the independence and noncollinearity of relevant variables. However, these assumptions do not sit well with the accounts of politics that have increasingly come to the fore in comparative analysis, which stress path dependence (Pierson 2000), or, as in this case, strategic interdependence among actors. Hall suggests that "process tracing" – the careful examination of paths of causation to see if outcomes are congruent with those predicted by theory – is a more appropriate way to test the validity of complex causal models.

I seek to build on insights from both the analytic narratives approach of Bates and his co-authors, and the historical institutionalist approach to process tracing advocated by Hall. On the one hand, I draw inspiration from the kinds of formal models emphasized in the analytic narratives approach.[3] On the other, I seek to use theory and evidence not only to provide an account of the relationships in a specific setting but also to draw conclusions that may potentially have wider applicability.[4]

Before so doing, however, it is appropriate to situate this book more precisely within existing debates in the field. In the remainder of this introductory chapter, I will set out in turn the main benefits and shortcomings of the political culture approach and current rational choice approaches to trust and cooperation. I will then go on to examine issues of trust and cooperation in the field of comparative political economy, with particular reference to the literature on industrial districts and other regionally based forms of economic cooperation. I then conclude by stating more precisely how this book contributes both to broader arguments about trust and cooperation and more specific debates about how these concepts explain outcomes in the political economy.

1.2 APPROACHES TO TRUST – POLITICAL CULTURE

Trust has been a central concept in comparative politics at least since the first wave of the behavioral revolution. This may in large part be traced back to the research agenda set out in Gabriel Almond and Sidney Verba's *The Civic Culture* (1963). Almond and Verba sought to

[3] As Bates et al. (1998a) note, the analytic narratives approach can be used together with non-rational-choice approaches; however, to date it has usually been identified with the rational choice paradigm.

[4] As Bates et al. (1998a) suggest, there are serious theoretical problems involved in reaching generalizations on the basis of game theoretic models that have been constructed to capture the specificities of a particular case. However, although the approach in this book borrows heavily from results in the game theoretic literature, its main argument is driven by mechanisms of equilibrium selection that are external to game theory, and that are thus less vulnerable to the problems that Bates et al. identify.

locate the sources of political success in the political culture of a country: those beliefs, norms, and values held by the country's citizens. Specifically, they argued that countries with "parochial" cultures, in which individuals failed to identify properly with their fellow citizens, instead relying on local or familial ties (Banfield 1958), were likely to do less well than countries with "subject" cultures, or (the ideal) "participant" cultures. In Almond and Verba's (1963, 284) view, this could in large part be traced back to social values stressing cooperation: "The degree to which cooperative interpersonal behavior is valued is directly related to the propensity to create political structures."

Trust rapidly became a key explanatory variable, if not *the* key variable for scholars of political culture. They argued that it explained the relative levels of development of different countries. Countries with a high level of diffuse interpersonal trust were likely to be politically and economically successful; those that did not were likely to do poorly. As described by Lucian Pye (1965, 22), "political cultures are built either upon the fundamental faith that it is possible to trust and work with fellowmen or upon the expectation that most people are to be distrusted and that strangers in particular are likely to be dangerous." This emphasis on trust as "one of the most basic ... attitudes" supporting mass democracy persists in contemporary work on political culture (Inglehart 1988, 1204). Ronald Inglehart, in a series of articles and monographs (1988; 1990; 1997; 1999), has sought to show that diffuse interpersonal trust remains a key explanatory variable for political science, and Robert Putnam's work, which is deeply influenced by the political culture tradition, also stresses trust as a vital explanatory factor.

Even if scholars of political culture see trust as a core concept, their account of its origins and effects has problematic microfoundations. Three problems stand out. First, the political culture account of diffuse interpersonal trust (and other cultural factors) rests on an insufficiently specified causal model. This has been a criticism of political culture accounts since Almond and Verba's original research. The concept of political culture had its beginnings in Parsonian sociology (Almond 1956), which notoriously employed a conceptually muddy and functionalist approach to the explanation of social institutions and practices.[5]

[5] For an intellectual genealogy that situates political culture firmly in the Parsonian tradition, while arguing that Almond and Verba got their concept of culture "exactly right," see Eckstein 1996.

Almond and Verba were robustly criticized for assuming that culture caused structure, rather than vice versa (Pateman 1980; see also Barry 1978). More recent versions of political culture theory have argued that the causal arrows run both ways, so that culture may affect structure/ and or politics, and vice versa (Inglehart 1990). In broad terms, this claim is inarguable, but without more specific – and testable – arguments as to how each relates to each, it is also rather unhelpful (Johnson 2003). Further, generalistic arguments of this sort may in practice shade into functionalist accounts of the origins of trust, as in Robert Putnam's *Making Democracy Work* (1993). For Putnam, civic values, networks, and norms (informal institutions) go together to make up social capital. He states that the norms of diffuse reciprocity that underpin cooperation and trust in society evolve "*because* they lower transaction costs and facilitate cooperation."[6] Arguments that seek to explain a social practice in terms of its broad social benefits, without providing clear causal mechanisms connecting the two, have little intellectual merit (Farrell and Knight 2003). In short, even while scholars of political culture have provided intriguing – and important – empirical evidence on topics such as the relationship between certain sets of values and material well-being in a society (Inglehart 1990), they have failed to provide a satisfactory account of where culture comes from and how it is related to other important factors, which in turn leads to implausibilities in their account of trust.

Second, the notion that trust is an "attitude" (Inglehart 1988), or that certain cultures have "dominant values [that] stress cooperative behavior" (Almond and Verba 1963, 284) rests on a deterministic, and conceptually flawed, account of what attitudes and values are (Johnson 2003). It is important to note that there are controversies among scholars of political culture about how best to conceptualize cultural variables. Scholars such as the late Harry Eckstein follow Parsons in suggesting that culture largely consists of learned patterns of orientation (Eckstein 1996; 1997). Under this account, culture is the product of socialization, and it plays a strong role in determining individuals' behavior. The general problems with accounts of this sort are well known (Granovetter 1985; Johnson 2003). However, they also have specific consequences for the understanding of trust.

[6] Putnam 1993, 172, emphasis added.

They suggest that diffuse interpersonal trust is something that is learned rather than considered, so that there is limited (if any) scope for individual agency or decision making in explaining it.

But there are scholars who view political culture as playing a less overwhelmingly determinative role. Both Elkins and Simeon (1979) and Barnes (1988) argue that culture does not define the individual so much as it provides her with an understanding of the assumptions that are shared in her society; she need not agree with a belief or orientation in order to recognize it as an important facet of her culture.[7] This second account of culture provides much greater scope for human agency and is at least partly compatible with a broadly rational choice perspective on cultural factors.[8] Unfortunately, however, it is Eckstein's rather less supple account of political culture that has predominated in empirical work. Ronald Inglehart, for example, applies Eckstein's concepts in his cross-national work on culture (Inglehart 1990), with direct consequences for his understanding of interpersonal trust, which he sees as resulting from very broad social forces (such as national history) rather than from the interplay between individual actors.

Finally, most work in the political culture tradition applies a particular set of tools to the understanding of culture: the statistical analysis of survey data. Even if one is not so profound a skeptic of the ability of survey methods to uncover cultural patterns as James Johnson (2003), one may still note that broadly framed questions, such as those in the World Values Survey, provide a doubtful proxy for how trust and cooperation actually work in a society (Hardin 2002). Surveys typically ask respondents whether or not they believe that others in their society can or cannot be trusted as a general rule. When patterns of trust or distrust involve specific individuals and specific matters (as they usually do), questions that seek to ascertain levels of generalized, impersonal trust are unlikely to provide helpful insights into actual motivations or behavior. Thus, it is at best unclear whether the survey data used to study trust is a good proxy for how individuals trust (or do not trust) each other.

[7] Johnson (2003), argues that this analytic "distance" between culture and individual personality traits is necessary if culture is properly to be understood.

[9] See Johnson 2002; for an empirical application, see Farrell 2003.

All of these problems suggest that political culture approaches to trust, at least as they are commonly construed, have problematic microfoundations. They rest on imprecise or functionalist causal models, apply a Parsonian theory of culture that provides little scope for individual agency, and use survey data that provides a poor proxy for actual trust and cooperation. A more successful account of trust and cooperation should rest on secure microfoundations, such as those provided by rational choice theory.[9] However, as discussed in the following section, rational choice theory faces its own theoretical problems in describing trust.

1.3 TRUST AND RATIONAL CHOICE THEORY

Rational choice approaches to politics, like political culture, take the problem of cooperation as a central concern. Noncooperative game theory, despite its name, provides a powerful set of tools for understanding the circumstances under which individuals will or will not cooperate with each other. What is less clear, though, is how well rational choice approaches can explain trust, which typically involves complex and personal relationships that are difficult to model using a game-tree or similar methodology.

Three approaches to the understanding of cooperation predominate in rational choice theory. First, scholars such as Oliver Williamson treat the decision of whether to cooperate or not with others as being for the most part an exercise in parametric decision making (Williamson 1975; 1985). Williamson suggests that the decision of a firm to produce internally or to outsource (cooperate with another firm) in a given instance, depends on the cost of the transaction at hand, which in large part is a function of the potential for opportunism. The risks of opportunism, however, primarily adhere to the type of transaction,

[9] I note that rational choice is not the only theory with strong microfoundations. There is an increasingly important literature in actor-centered sociology, not all of which, by any means, adopts a rational choice perspective. See, for example, Fligstein 1996. While some of this literature is less sensitive to considerations of power and interest than the perspective set out in this book, it in turn may capture aspects of cooperation that are difficult to encompass in a broad rational choice framework.

rather than the type of partner.[10] Unsurprisingly, Williamson has little patience for theories of trust, arguing that they have little purchase in the realm of economic decision making, where interests are unambiguous (Williamson 1993). Thus, Williamson argues that trust is unnecessary to the explanation of economic cooperation; instead, cooperation flows more or less directly from the interests of economic actors.[11]

Second, there is a burgeoning literature that treats trust and cooperation as features of games. Work in this vein ranges from relatively simple treatments such as Partha Dasgupta's "Dishonest Salesman" game (1988) to the sophisticated use of game theory to model bilateral cooperation within a large group of agents.[12] In the latter body of work, infinitely repeated games are used to model institutions that "promote the trust necessary for efficient exchange" (Milgrom, North, and Weingast 1990, 1) Institutions are usually represented as subgame perfect equilibria, in which no actor has an incentive to deviate from her strategy.[13] Given appropriate parameter conditions, and the existence of a specific institution, actors will have no incentive to behave in an untrustworthy fashion, so that trust and cooperation can be maintained among quite large groups of actors. Institutions, under this approach, may involve quite complex sets of social relationships and information exchange, in which, for example, one actor may serve as a central clearinghouse for information.

[10] This is perhaps one of the reasons for the conceptual ambiguities in Williamson's efforts to update his original model (1975), which treated only of markets and hierarchy, to include networks as a third form of governance (1985). In including networks, Williamson attempts to incorporate long-term relationships between firms, without fully examining the implications of strategic action for these relationships, thus creating a set of arguments which, while powerfully suggestive, have important lacunae.

[11] Williamson (1993) acknowledges a role for "institutional trust," but his remarks on what this might involve are not entirely transparent.

[12] See, especially, Milgrom, North, and Weingast 1990, Greif 1994, and Greif, Milgrom, and Weingast 1995. For a theoretical *summa* of this approach, see Calvert 1995b.

[13] There is some disagreement among scholars as to the precise definition of an institution. Douglass North (1990) and Jack Knight (1992), both define institutions as sets of rules and distinguish them from organizations, which are collectivities of actors. Calvert (1995b) starts from game theory and sees both institutions and organizations as equilibria, albeit differing in their levels of complexity. I follow North and Knight's definition, which I suggest provides more specificity and thus greater purchase on the root causes of cooperation.

Less formal accounts of institutions and trust may be found in that branch of the New Institutional Economics (North 1990; North and Weingast 1995) that emphasizes how institutions may underpin credible commitments. In this argument, institutions may serve as an important technology of commitment. Actors may subject themselves to institutions when they seek to demonstrate to others that they may be trusted to behave honestly in situations where those others might reasonably suspect that these individuals would otherwise behave opportunistically.[14] In both this literature and in the game theoretic literature discussed previously, trustworthiness flows more or less ineluctably from certain institutional arrangements. Institutions both serve to underpin trustworthiness (by giving actors the incentive to behave in a trustworthy fashion) and to communicate information about the trustworthiness or untrustworthiness of others.[15]

Finally, a recent and important literature seeks to build a theory of personal trust on the basis of considered interests. Russell Hardin's "encapsulated interest" account provides a three-part account of trust in which one actor trusts another when she knows that the latter actor's interests encapsulate her own with regard to the relevant matter.[16] As Hardin emphasizes, this is a relational account of trust – actor interests will most likely encapsulate each other when the actors are engaged in a long-term relationship, which might deteriorate or be lost if one of them behaves in an untrustworthy fashion. Hardin also claims that broader social institutions may affect trust between actors, although his argument about the relationship between social institutions and individual relationships that involve trust is not fully spelled out. In Hardin's view, it is unlikely that one will trust others whom one does not know directly, such as government officials. Margaret Levi (1998), who has also sought to develop the encapsulated interest account, disagrees (although see Cook, Hardin, and Levi 2005). She suggests that individuals may trust government officials and others when the latter embed themselves in institutions that give them the right incentives.

14 See Farrell (2004) for more extensive discussion of the relationship between credible commitments and trust.

15 See, for example, the discussion in Calvert (1995b), although note that in his model, behaving in an untrustworthy fashion is off the path of play.

16 The encapsulated interest account is discussed in greater length in Chapter 2.

The argument that I develop is a rational choice one: I construct an argument about trust, cooperation, and institutions on the basis of arguments about actors who pursue goals that reflect their individual economic interests.[17] However, there are important gaps in the rational choice literature as it now stands. First, much of the literature on institutions, trust, and cooperation is implicitly or explicitly functionalist. The argument that institutions may support cooperation is all too frequently conflated with the argument that institutions come into being "in order to" support cooperation. Functionalism is especially endemic in the game theoretic literature on institutions. Although game theorists have enjoyed remarkable success in modeling quite complex social institutions as equilibria in infinitely repeated games, they face the problem of the "folk theorem," which suggests that there are vast amounts of potential equilibria in such games, each of which is equally plausible from a theoretic point of view. Thus, game theorists may use game theory to say that an equilibrium is possible under certain parameters, but they very often have to turn to external mechanisms in order to explain why one equilibrium rather than another has prevailed. All too often, however, they merely show that an equilibrium is possible and would further a broad social goal without further discussing the specific mechanisms through which this equilibrium rather than another was arrived at.[18] Many articles on trust and cooperation thus model an institution that supports cooperation among a group of actors and assume that the institution came into being in order to support cooperation, rather than investigating the various causal mechanisms that may be implicated in institutional creation or change.

Furthermore, there is a disjuncture between institutionalist approaches, which are overly determinist, and relational accounts of trust, which have difficulty in explaining how institutions affect trust. Despite some important initial contributions (Kreps 1990a; Miller 1992; Levi 1997; 1998; Ensminger 2003), we still do not have a fully developed theory of how institutions may *affect* trust and cooperation among actors without fully *determining* their interests.

[17] I note that the encapsulated interest account is quite compatible with more expansive perspectives on rationality; see Hardin 2002.

[18] I discuss this problem at greater length in Chapter 2.

The institutionalist literature provides an overly simplistic and deterministic account of trust and cooperation. For example, Milgrom, North, and Weingast (1990) demonstrate the existence of an institutional equilibrium under which dishonest behavior lies off the path of play. It is not in any player's interests to cheat, given certain parameters and expectations about how other players will respond. This captures an important effect of institutions. However, it is unhelpful to conflate this effect with trust. Trust is usually seen as involving social relationships that are personal, at least in part, and matters in which there is a real (if unquantifiable) risk of default. Game theoretic models of institutions have extreme difficulty in modeling such relationships and the kinds of expectation generated by these relationships. To the extent that cooperative relations are a simple function of the incentives created by institutions, Williamson's criticism stands. It is a misnomer to describe expectations over straightforward institutional compliance as trust.[19]

The encapsulated interest account of trust provides a nuanced account of trust among actors that is sensitive both to the uncertainties involved in trust and to its specifically relational nature. However, precisely because it treats trust as a relational concept, it has difficulties in delineating how institutions may affect it. Insofar as trust involves personalized relationships among specific individuals, it is at least at one remove from the effects of institutions.

Therefore, rational choice approaches to trust offer both a useful theory of how institutions may induce cooperation and a helpful microlevel account of trust in personal relations. However, it still has difficulty in creating bridges between these accounts, that is, in showing how institutions may affect trust in relationships among individuals, as opposed to merely inducing determinate expectations over cooperation.

1.4 TRUST AND POLITICAL ECONOMY

Controversies about trust and cooperation have become increasingly important in the field of comparative political economy as that field has

[19] Note, however, that I do not endorse Williamson's corollary argument that trust is unconsidered and is confined to purely personal relationships.

broadened in scope and ambition. Increasingly, comparative political economy is concerned not only with the sources and consequences of government economic policy across nations but also with the broader patterns of cooperation and competition that characterize different models of capitalism (Crouch and Streeck 1997); Hollingsworth and Boyer 1997; Hall and Soskice 2001). Comparative political economy has also seen an increasing degree of agreement on basic principles, as historical institutionalists and rational choice institutionalists have begun to use models and arguments that rest on similar microfoundations.[20] Historical institutionalists and rational choice scholars now not only agree that institutions are important but they also increasingly agree on *why* they are important, and what they are. Peter Hall and David Soskice, for example, argue on behalf of an

> approach to comparative capitalism [that] emphasizes the presence of institutions providing capacities for the exchange of information, monitoring, and the sanctioning of defections relevant to cooperative behavior among firms and other actors.[21]

Few rational choice scholars, if any, would disagree with this set of priorities. Thus, for Hall and Soskice, as for other comparative political economists, the relationship between institutions and cooperation is vital to how economies work.

However, some important problems remain in Hall and Soskice's account, as in other work in the literature on comparative capitalism. Specifically, the relationship between micro- and macrolevel phenomena is poorly spelled out. Hall and Soskice argue that their account provides a firm-centered approach of the economy and compare economies according to how firms resolve their coordination problems.

[20] This development should not be seen as the "conquest" of comparative political economy by rational choice. While historical institutionalists increasingly share common presuppositions with rational choice scholars, they continue to be more open to dialogue with ideationalists (Hall and Taylor 1996; Thelen 1999; Mahoney 2003). Furthermore, historical institutionalists' development of path dependence arguments has given them a new set of tools that are allowing them to sharpen their arguments about the precise circumstances under which history "matters." See Thelen 1999; 2004, Pierson 2000; 2004, Mahoney 2003, and Crouch and Farrell 2004. I discuss path dependence theory further in Chapter 2.

[21] Hall and Soskice 2001, 11. Indeed, Hall and Soskice explicitly "construe the key relationships in the political economy in game-theoretic terms and focus on the kinds of institutions that alter the outcomes of strategic interactions" (5).

They distinguish between "liberal market economies," in which firms primarily rely on market mechanisms to coordinate, and "coordinated market economies," where they rely on nonmarket mechanisms, including incomplete contracting. In this perspective, institutions are important because of "the support they provide for the various types of relationships that firms use to resolve coordination problems" (Hall and Soskice 2001, 9). Hall and Soskice do not seek to explain how nations acquire these institutions in the first place. However, they suggest that nations with certain institutions in one sphere of the economy will seek to develop "complementary" institutions – institutions that involve complementary forms of coordination – in other spheres. Thus, one may expect that economies will be governed by relatively coherent institutional ensembles at the national level.

Hall and Soskice's account has some important advantages over much existing work: it demonstrates how institutions can affect cooperation between economic actors without fully determining the form that cooperation takes by providing greater support for some types of relationship than for others. However, it also has some important lacunae. Hall and Soskice state that their account (1) relies on a strategic account of individual interaction, and (2) only seeks to explain how institutional structures affect coordination among firms, and not the origins of the institutions themselves. To the extent that the authors follow their own prescriptions, they fall into an implicit functionalism, in which strategic interaction is only important insofar as it creates problems that need to be "solved" through different institutional structures. In short, their approach assumes that institutions' role is to improve coordination and has little to say about the strategic problems involved in the process of institutional creation itself. This is most obvious in Hall and Soskice's account of institutional complementarities, which is a cornerstone of their larger intellectual project. They suggest that certain kinds of institutions tend to be found together in an economy because "the presence (or efficiency) of one increases the returns from (or efficiency of) the other" (Hall and Soskice 2001, 17). In the absence of specific propositions as to how strategic interaction among actors leads them to seek efficiency gains, this is a functionalist claim. More recent work by Hall on his own (2005), and in collaboration with Kathleen Thelen (Hall and Thelen 2007), goes some distance toward articulating an account of institutional change that pays more attention to distributional struggle. However, precisely because of its eschewal of functionalism, it sits uneasily with Hall and Soskice's macrolevel framework. In short, Hall and Soskice seek to introduce a strategic element to the understanding of cooperation in the comparative political economy, but they do not go far enough. Strategic interaction exists in their model as a problem to be resolved through appropriate institutions, rather than an underlying feature of the processes that create institutions in the first place.[22]

1.5 ARTICULATING AN ALTERNATIVE ACCOUNT OF TRUST IN THE ECONOMY

In the previous sections I have discussed existing approaches to the explanation of trust and cooperation in the political economy, and I have suggested that these accounts have some rather important shortcomings. I identify two key problems. First, there is a strong tendency toward functionalism in most current approaches to the understanding of trust. Scholars tend to assume, explicitly or implicitly, that the social structures that support trust between actors come into being *because* they support trust between actors. This is as true of much rational choice or "varieties of capitalism" work examining how institutions allow actors to cooperate with each other, as it is of political culture scholarship, which assumes that political culture exists because it fulfils the broader functional needs of a society. In specific instances, it may well be that the origins of social structures are explained by the social needs that they meet, but this should be a subject for empirical investigation, rather than a starting assumption. Not only may functionalist explanations be wrong (as they very often are), but they usually commingle the *explanans* and the *explanandum* so as to muddy causal relationships. Thus, a satisfactory account of the relationship between institutions and trust should not rest on functionalist assumptions, unless there is strong independent evidence

[22] I note that Hall and Soskice's contribution may be criticized from other vantage points. I note also that the strategic perspective that I proffer here has its own theoretical blind spots. For an alternative approach, see Crouch and Farrell 2004. However, the point that I wish to stress is that Hall and Soskice promise a strategic account of comparative political economy that applies insights from recent game theoretic work on institutions, but they fail to deliver fully.

that functional needs are in fact the main driving force in the relationship at hand. Instead, it should be able to show how social structure may affect trust relations, without necessarily arising from the need to underpin these relations.

Second, current accounts have great difficulty in describing the relationship between the macro- and microlevels, that is, between the structures that influence trust relationships and those trust relationships themselves. Political culture accounts have great difficulty in speaking to the microlevel on which individuals trust or do not trust each other with regard to specific matters, while rational choice accounts have problems in linking macro- to micro-level without reducing trust to quite narrow institution-induced expectations. A satisfactory account should thus be able to show how the macro- and microlevels are related, without reducing the one to a mere effect of the other. As I have noted, current work in the literature on "varieties of capitalism" goes some way toward doing this, but on the basis of a causal model that has quite important gaps and ambiguities.

I propose an account of the "political economy of trust," which, I hope, meets both of these criteria. It does so on the basis of rational choice theory and, more specifically, the rational choice theory of institutions. This means that it has some clear limits. Institutions are not the only form of social structure that is relevant to trust relations. As sociologists have been at pains to illustrate, social networks may have quite important effects on individual relations, which are by no means fully reducible to institutions.[23] However, the converse is also true, and network accounts of politics face enormous difficulties in bridging the micro- and macrolevels without invoking external mechanisms.[24] Furthermore, precisely because my theory of trust is a rationalist one (and one that privileges material interests at that) it

[23] For discussion of networks, see Wellman and Berkowitz 1988; White 1988; Padgett and Ansell 1993; Locke 1995.

[24] The statistical analysis of social networks becomes exponentially more difficult as the size of the network increases – even with modern computing technology, it is impracticable to analyze relationships in very large networks. A recent literature has begun to employ different tools to explore the consequences of network topography in networks with very large number of nodes; see Farrell and Drezner (2008) for an application, and also Newman Barbási, and Watts (2006) for an excellent overview. However, this body of work has only scant relevance to the issues of trust and cooperation that this book examines.

can only explain forms of trust that rest on one kind or another of self-interest. Thus, it cannot explain *trust beyond interest*, that is, the grounded expectation that actors will be trustworthy in circumstances where it is not in their best interest to be so. I do not seek to deny that such trust (and trustworthiness) exist. Indeed, I acknowledge that I cannot explain it using the arguments laid out in this book. Instead, I seek to show how an interest-based account of trust can accommodate institutional effects that go beyond the kinds of determinate effects on expectations provided by game theoretic models and the like.

These caveats aside, there is considerable reason to believe that an institutionalist account of trust may provide important insights into how trust comes about, and how it affects social relations. As Jack Knight puts it:

> the concern with such issues as social capital, trust and civility basically involves an analysis of the types of informal institutions that exist in a society ... to understand the emergence and maintenance of social capital, trust and community in a society, we must understand the evolution of these informal rules.[25]

By coming to a better understanding of the consequences of institutions, we may better appreciate the sources of trust and cooperation. Furthermore, even if some forms of trust are not based on expectations over self-interest, many are. It is furthermore plausible that self-interest-based forms of trust will play a much more direct role in areas such as economic relations, where material interests are both obvious and important.

I do differ substantially from many other scholars in my account of the relationship between institutions, trust, and cooperation. Rather than positing functional complementarities and feedback loops (Hall and Soskice 2001), I suggest that institutional change is driven by self-centered actors' efforts to reap distributional benefits (Knight 1992; Farrell and Knight 2003). Thus, I seek specifically to distinguish between the *sources* of institutions – which I argue, following Knight (1992), are struggles over how benefits are distributed – and the *consequences* of institutions (observable patterns of trust and cooperation at least in some instances). Furthermore, I hope to provide a more

[25] Knight 1998, 754.

complex and nuanced account of the relationship between institutions and trust than provided by the extant literature, showing how trust is related to institutions but cannot simply be reduced to institution-induced expectations.[26]

In the social sciences, perhaps the most important test of a theoretical framework is whether it gives a good explanation of empirical reality. Without seeking to engage directly in ongoing and complex debates about the epistemology of the social sciences, I suggest that any framework should meet two important criteria. First, and most obviously, are its predictions supported by the available empirical evidence? If not, then the theory may still have real intellectual value (it may involve substantial methodological or theoretical advances, or it may provide interesting counterfactuals), but it is likely to be of limited practical relevance to our understanding of politics. Second, does the theory make nonobvious predictions? Does it advance our understanding of empirical phenomena in important and nontrivial ways? If not, its contribution to our scientific knowledge is again likely to be limited: proving the obvious only infrequently provides significant advances in understanding.

I will show that an institutional account of trust and cooperation does indeed seem to be supported by empirical evidence gathered from important cases privileged by the existing literature. Furthermore, it sheds light on an unsolved empirical puzzle: the sources of inter-firm cooperation within industrial districts. Inter-firm cooperation is one of the most important forms of economic activity within advanced industrial societies – different patterns of cooperation are associated with different "varieties of capitalism" (Hall and Soskice 2001). Industrial districts provide an important test case for theories of trust. Indeed, debates about industrial districts and regional economies have played an important role in the development of modern comparative political economy as a field of study. However, these districts are difficult to explain using the theoretical frameworks that have dominated the field. The "classic" industrial districts involve forms of inter-firm cooperation that do not sit well with typical transaction cost accounts of the circumstances under which actors do or do not

[26] This account builds upon joint work with Jack Knight; see Farrell and Knight (unpublished).

cooperate (Trigilia 1998). The political culture perspective faces even greater difficulties in explaining trust and cooperation in industrial districts. The *loci classici* of the phenomenon are located in the center and northeast of Italy, a country that has frequently served as political culture theorists' prime example of a low-trust country (Almond and Verba 1963; LaPalombara 1965; Inglehart 1990). How is it that such extensive forms of trust are possible in a country where there is scanty "culture of trust" to support it?[27]

Chapter 6 begins from a different yet related question. Members of the Sicilian mafia live within a dense set of mutual understandings and expectations, which are structured by informal institutions. Some of these informal institutions (such as the requirement that mafiosi always speak the truth to each other) seem on the face likely to support trust and cooperation, yet in practice they appear not to do so. Why is this so?

An institutionalist account of trust provides us with important insights as to how economic actors may come to trust each other enough to engage in the extensive forms of cooperation that are found in industrial districts. It shows us how institutions may support cooperative relationships among actors that may enable the radical disintegration of the production process. It also provides a more nuanced and specific account of trust at the microlevel than prevailing theories. While I do not suggest that my account provides a *complete* theory of trust (humans are neither fully rational nor entirely motivated by self-interest, nor only guided by institutions), it does provide a *sufficient* account to explain how trust may arise in apparently unpromising circumstances.

In addition, an institutionalist theory of trust provides us with important nonobvious insights. It provides an explanation of key empirical puzzles. Why is it that certain kinds of trust are common in societies that have weak formal institutions and little generalized trust, but they are in societies where individuals, at first glance, face much less serious problems of trust and cooperation? Why is it that informal institutions seem to support trust under some circumstances and to perpetuate distrust under others? Through the careful

[27] As I go on to discuss, regional differences within Italy are insufficient to explain this (Putnam 1993).

comparison of case studies involving different kinds of institutions, I provide answers to these questions.

I offer an alternative account of the political economy of trust, starting from the bottom up. Chapter 2 sets out an extensive theory of the relationship between institutions, trust, and cooperation. It returns to debates over institutions and debates over trust in greater detail, setting out a theory of how institutions may affect trust without reducing trust to simple expectations over institutional compliance.

Chapter 3 situates the main empirical discussion by providing brief accounts of the industrial district phenomenon and of the two industrial districts that are examined in the bulk of this book. Its intention is to offer readers without previous grounding in this literature an account of what industrial districts are, why they are important, and why the specific industrial districts examined in this book may usefully be employed as proxies for a more general phenomenon. It concludes by arguing for the benefits of a rational choice account of trust, which focuses on relatively stable individual interests, in explaining why actors trust each other or do not trust each other, and why they cooperate or do not cooperate in industrial districts.

In the subsequent empirical discussion, I build up an account of trust, institutions and cooperation. In Chapter 4 I examine whether or not institutions affect the ways in which actors trust or do not trust each other, and how best to capture the putative effects of institutions by examining institutions comparatively. If institutions affect the ways in which actors trust or do not trust each other, one may reasonably expect that different institutional settings will be associated with differences in trust, and thus in cooperative outcomes. In particular, I argue that one may expect to see systematic differences between settings where formal institutions predominate and settings where informal institutions are more important.

Chapter 5 considers the sources of institutional creation and change. In order to avoid the rather diffuse functionalism that frequently characterize arguments about trust and trust creation, I specifically eschew an account that explains institutional change in terms of its functional benefits, instead assessing an account in which institutions instantiate social beliefs that reflect power relations (Knight 1992; Farrell and Knight 2003). I use this account to structure an analytic narrative that provides an account of informal institutional changes in one of the two cases (the other, because it involves formal rather than informal institutions, does not provide suitable data for this purpose).

Chapter 6 turns to a quite different context – trust relations among members of the Sicilian Mafia – in order to investigate the circumstances under which informal institutions may perpetuate distrust. It examines how the informal rules governing life within the Mafia set generalized expectations over behavior and distort information flows so as to exacerbate persistent problems of distrust among mafiosi.

The conclusion discusses the applicability of my claims to debates about the relationship between trust and self-interest, between trust and institutions, and also how trust relations may change as a result of institutional change.

2

A Theory of Institutions and Trust

2.1 INTRODUCTION

I have discussed the gaps in the existing literature on the relationship between institutions and trust. But there is a broader research agenda, one that might help us arrive at answers to some of the key questions of comparative politics. What implications do informal and formal institutions have for the workings of the economy? How can we best understand the sources – and consequences – of trust and cooperation? Bringing these together, what effects do institutions have on the way that individuals trust (or do not trust) and cooperate with (or do not cooperate with) each other?

One of the main reasons that we do not have good empirical accounts of the relationship between institutions and trust is because of the weakness of our underlying theories of this relationship (Johnson 2003). As I have noted, some treatments of the relationship (including many prominent examples of the political culture approach) tend toward excessive vagueness, others (including many game theoretic accounts) are reductionist, treating trust merely as a highly determinate set of expectations. Thus, a more coherent theoretical account of the relationship between institutions and trust is a necessary first step toward a better empirical understanding of the relationship.

The most sophisticated accounts of trust that we currently possess have little room for institutions. Instead they stress personal relationships as the key source of information underlying trusting beliefs.

Our theories of institutions are even less able to help explicate trust; dominant theories in rational choice effectively elide trust out of existence, concentrating on simple institutional compliance. How can we arrive at a theory of trust that on the one hand incorporates the effects of institutions, but on the other does not simply reduce trust to institution-induced expectations?

Here I present an alternative account of trust and cooperation that directly incorporates the influence of institutions. This account avoids functionalism, emphasizing how *power relations*, rather than the functional need for cooperation, lead to institutional change, and thus help shape trust and cooperation among individuals. It furthermore shows how different *kinds* of institutions are likely to have quite different consequences for trust and cooperation. Much recent work in the social sciences has sought to distinguish between informal and formal institutions, frequently using noncooperative game theory to model institutional enforcement. Informal and formal institutions are likely to have important consequences for trust and cooperation that go beyond their simple enforcement characteristics.

Let us examine Russell Hardin's claims about the beliefs underlying trust and their origins. Hardin's encapsulated interest approach arguably provides the most sophisticated current account of how trust works, and thus it provides an excellent beginning point for thinking about trust. However, because it stresses personal relations, it does not readily lend itself to the specific task of theorizing the relationship between institutions and trust. The chapter argues that by considering trust as a set of expectations about the trustworthiness of others in imperfectly defined future situations, we may begin to understand how institutions can provide information about other types of actors, and how they are likely to behave in such situations, thus contributing to trust or distrust among these actors. The next two sections examine theories of institutional *origins* and *consequences*, arguing that dominant approaches to the understanding of both present major conceptual difficulties to our understanding of how institutions might affect trust. Instead, this book builds on the work of those (e.g., Moe 1987; 2005); Knight 1992; 1995; Farrell and Knight 2003) who examine the ways in which power relations are instantiated in institutions, arguing that this provides a more appropriate theory of institutional origins and change than either functionalist or path-dependence accounts. Its account of institutions' effects seeks to provide an alternative understanding to the dominant account in the existing literature: one more open to the complex expectations that make up trust. It argues that institutions are best conceived of as mechanisms of equilibrium selection that make some equilibria more likely, and other equilibria less likely, to be chosen by the relevant actors. Finally, theories of institutions and trust are brought together – the concept of trust should only be invoked in circumstances where institutional rules do not have fully determinate effects. It discusses both (1) how change in power relations is likely to affect institutions, and thus trust, and (2) how variation in institutions is likely to be associated with variation in the level and kind of trust among actors.

2.2 THE ENCAPSULATED INTEREST ACCOUNT OF TRUST

What is trust? Even compared with other social scientific concepts, trust is protean – different authors use the term to refer to very different things. I follow Russell Hardin (2002) in treating trust as a belief or set of beliefs of a quite particular kind. Trust is best considered as a set of beliefs or expectations "about the cooperative intentions of others." When I trust you, I believe that you are likely to cooperate in the future. Thus, while trust and cooperation are linked to each other, they are conceptually quite distinct. The former is an expectation, the latter an action. More specifically, cooperation involves costly action; in Randall Calvert's description, it means that "an individual engages in some act whose immediate consequences for that individual, regardless of what others do at that moment are negative, but which generates some positive benefits for some other individual (and usually for several others in a group)" (Calvert 1995a, 224; see also Williams 1988). To say that I trust someone is to say that I believe that this person will cooperate by engaging in a costly action to my benefit should this prove necessary.

However, not all expectations regarding others' propensity to cooperate qualify as trust. When actors have good reason to be *certain* that others will cooperate, these expectations are better described as confidence than as trust. In Diego Gambetta's words:

> The condition of ignorance or uncertainty about other people's behaviour is central to the notion of trust. It is related to the limits of our capacity ever to

> achieve a full knowledge of others, their motives, their responses ... Trust is also related to the fact that agents have a degree of freedom to disappoint our expectations. For trust to be relevant, there must be the possibility of *exit*, betrayal, defection. If other people's actions were heavily constrained, the role of trust in governing our decisions would be proportionately smaller, for the more limited people's freedom, the more restricted the field of actions in which we are required to guess *ex ante* the probability of their performing them. (Gambetta 1988, 218–19)

More specifically, as Edward Lorenz (1999) argues, trust involves the expectation that another will cooperate *ex post* in circumstances that cannot readily be anticipated *ex ante*. This is key to the distinction between confidence and trust. When I know that another will behave honestly in a predetermined and well-anticipated situation, this knowledge is better described as confidence than trust. Trust is only really implicated where there is the possibility of an unanticipated, or only imperfectly anticipated, situation arising, where I am obliged to determine, on the basis of my existing knowledge and beliefs, whether someone is trustworthy or not.

Russell Hardin's "encapsulated interest" account of trust stresses personal relations rather than institutions as a key source of grounded expectations that may underpin trust. It nonetheless provides a valuable way to begin to think about trust – and then, in a separate step, to connect it to institutions. Hardin argues that expectations about the willingness of others to cooperate are grounded in perceptions of the interests that those others have in behaving cooperatively because of their particular regard for the interests of the trusting party.[1] More precisely, I trust you to the extent that your interests encapsulate mine with regard to a specific matter. Thus, Hardin's account makes three key claims as to what trust involves.

First, it claims that *trust* involves considered expectations over *trustworthiness*. Frequently the two concepts are elided, so that trust and trustworthiness are treated interchangeably; Hardin argues that they are very different from each other. Trust refers to the considered expectations of the trusting party that the trusted actor will

[1] I do not engage in an extended discussion of the encapsulated interest account here, only mentioning those points necessary to my argument. An extended discussion of the encapsulated interest account can be found in Hardin (2002). See also Levi (1998).

behave in a trustworthy fashion, that is, will take her interests into account. Trustworthiness refers to the willingness of the trusted party to honor the trust placed in her. While the two are obviously related, they are distinct from each other and need not coincide. If I am the trusting party, I may be mistaken in my expectations and trust you, even though you are not in fact trustworthy. Thus, as already discussed, trust always involves some degree of uncertainty, and cooperation based on trust necessarily involves risk. I may be more likely to make a costly cooperative move myself if I trust the other individual(s) involved to reciprocate by cooperating later, but I can never eliminate the risk that I am mistaken in trusting this/these other(s).[2]

Second, it implies that trust pertains to relationships with *specific others over specific matters.* Hardin is skeptical about the existence of generalized trust. As he says, the wide body of survey research that seeks to measure whether respondents trust other people in their society in general begs a number of very important questions. The kinds of trust that these survey questions seek to measure are of questionable social relevance; they do not necessarily describe very well how individuals trust each other in real-life situations, where one does not make decisions over whether or not to trust a generalized other but rather a specific individual or group. For example, even if I live in a supposedly high-trust society, it is unlikely that I would trust a stranger who approached me on the street and asked for a loan of $100 (Hardin 2002). Instead, my degree of trust will vary with regard to both the person who is asking for my trust and the matter that they are asking me to trust them on. I am likely to trust my new colleague who asks me to loan her $10 to pay for lunch, although I may not trust her if she asks me to loan her $100,000. I may not trust a random stranger even with a loan of $10. (Even though I may give it to them, it is not in any expectation that the loan will be repaid.) In contrast, I may trust a family member or a very close friend who asks me for a loan of $100,000, if I have it.

Thus, Hardin argues that the degree to which I trust others will vary according to my beliefs over their interest in behaving in a trustworthy fashion. Where I believe that another person has a strong interest in behaving in a trustworthy manner with regard to

[2] See Gambetta 1988.

a particular matter, I will be likely to trust them with regard to that matter. I may not trust them with regard to other matters, where I do not believe that they have an interest in behaving trustworthily, and I may not trust other actors (who may not have the same interests in being trustworthy) with regard to the same matter. As a consequence, Hardin defines trust as a three-part relationship. Its three components are the person doing the trusting, the person who is trusted, and the matter over which the latter is being trusted. Actor *a* may trust Actor *b* with regard to matter *c*. We may see variation in trust associated with variation in each of the three parts. Actor *a* may not trust Actor *b* with regard to another matter, *d*. Actor *a* may trust *b* with regard to *c* but not trust a different actor, *e*, with regard to *c*. Finally, another actor, *f*, might not trust Actor *b* with regard to matter *c*.

Third, Hardin argues that trust is an expectation, but one of a quite particular kind. It involves *a*'s expectations on the basis of *b*'s specific interests with regard to her, that is, the extent to which *b* will take *a*'s interests into account *because* they are *a*'s interests. To reiterate, *a* will trust *b* to the extent that she believes that *a*'s interests encapsulate her own with regard to the matter at hand. Thus, in Hardin's account, trust is a relational concept – the existence of a relationship between *a* and *b* gives *b* an interest in behaving in a trustworthy fashion, and thus gives *a* good reason to trust her. In Hardin's argument, this means that a wide set of social relationships involving expectations do not involve trust. When I am driving my car, I may expect others to drive on the same side of the road as I do. However, these expectations do not involve trust as such, since the other drivers do not have any specific relationship with me and thus have no interest stemming from that relationship that would prompt them to act in a trustworthy fashion. Instead, the simple knowledge that a coordination rule exists is sufficient to create convergence in interests, without any need for specific trust relationships among individual drivers. Hardin further suggests that this means that it may be problematic to apply the concept of trust to situations where our reasons for "trust" are not grounded in personal relationships, such as, for example, our relations with government officials. While trust is possible in principle in such relationships, we cannot expect most citizens to have the substantial knowledge of these officials' incentives that would be necessary for genuine trust to apply.[3]

Hardin's arguments present a serious quandary for scholars who wish to claim that trust is affected by institutions in important ways. Hardin certainly acknowledges that trust may be indirectly affected by institutions; at the very least, institutions may help to alleviate distrust by taking certain possible actions off the table. But if Hardin's claims stick, it is difficult and perhaps impossible to see how any strong version of trust could be grounded in institutions. As I argue at length below, institutions involve a higher level of aggregation and a more abstract form of knowledge than the personal relationships that Hardin privileges. In particular, they involve relationships between general classes of actors or, alternatively, means for transmitting information about individuals' reputations, rather than the particular and sometimes intimate relationships that we may have with others whom we know directly. Hardin suggests that the kinds of knowledge that we may gather from less personal relationships are too thin to support trust. Instead, we require relationships among actors whose personal dispositions are known to each other (Williams 1988) and who are engaged in ongoing relationships that potentially give them an interest in behaving trustworthily. Institutions on their own do not provide sufficient grounding for trust. In the remainder of this book, I use Hardin's arguments about the relationship between trust and trustworthiness, while maintaining *contra* Hardin that institutions *can* provide sufficient knowledge to ground trusting relationships. To develop this argument, however, it is necessary first to discuss how institutions come into being and what consequences they have.

2.3 COOPERATION, INSTITUTIONAL ORIGINS, AND INSTITUTIONAL CHANGE

If, as Hardin suggests, we have difficulty in building a theory of trust upwards from personal relations to encompass institutions, might it

[3] See Hardin 1998, 16: "If our notion of trust comes from understandings of individual behavior and character, the term may be entirely out of place in application to a nation, group, or institution. There may be ways to interpret the notion to apply it to such actors, but it is not likely to be prima facie applicable without interpretation."

be easier to build downwards from institutions, showing how institutions affect trust? Unfortunately, current theories of institutions have a difficult time in encompassing trust and cooperation. In part, this is because currently dominant theories have a tendency to explain too much by grounding their accounts of institutional emergence and change in the functional need to create cooperation and trust. The assumption that all institutions are created to maximize trust and cooperation makes it difficult to construct a useful theory of comparative variation in institutions (e.g., one that might help explain why trust is more prevalent in some institutional settings than others). Even more pertinently, it makes it hard to think about institutions independently from their consequences for trust and cooperation. This is largely because our theories of institutional *effects* are relatively crude; rational choice accounts in particular tend to treat institutions' effects on behavior as coterminous with institutional compliance.

Just as the term "trust" has many different meanings in the social sciences, so too does the term "institution." However, even though the "new institutionalism" reflects at least three, quite distinct, schools of thought, which originally had little contact with each other (Hall and Taylor 1996), the last several years have seen the beginnings of a rough consensus emerge about what institutions are and why they matter to politics (DiMaggio 1998; Katznelson and Weingast 2005). Both rational choice institutionalists and historical institutionalists agree that institutions are best thought of as rules or sets of rules that pattern human behavior. However, there is still some considerable disagreement among scholars as to where institutions come from and how they change over time. Many rational choice scholars argue that institutions come into being to serve functional needs, others (whether rational choice or historical institutionalists) disagree (Pierson 2000; Miller 2000).

This dispute is directly relevant to the relationship between institutions, trust, and cooperation precisely because the function that institutions are supposed to serve is to underpin cooperation (Weingast 2002). Much recent theoretic work on institutions explains the origins and forms of institutions specifically in terms of the ways in which they support cooperation, trust, or both. Oliver Williamson's (1975; 1985) transaction cost account of institutions is a case in point. Williamson (1985; 1991) argues that the "economic institutions of capitalism" come into being in order to mitigate transaction costs and most prominently in order to address the risks of opportunism associated with asset specificity. Hence, they allow economic cooperation to take place.[4] Thus, his argument draws a direct link between institutions and cooperation, which may take place either under hierarchy (within the firm) or through a variety of relational and nonrelational contracts (between firms). Robert O. Keohane's (1984) highly influential account of international politics draws on the new institutional economics to argue that international institutions come into being in order to underpin cooperation among states. Douglass North and Barry Weingast's (1995) account of the Glorious Revolution explains how "tying the king's hands" provided the enabling conditions for a cooperative equilibrium that was beneficial both to the king and to his debtors. When the king of England no longer had the power to renege on his debts, he ipso facto became more trustworthy to lenders.[5] Robert Putnam (1993) draws a direct link to trust as well as cooperation when he argues that the informal social institutions supporting trust in Northern Italy evolved in order to underpin cooperation. Milgrom, North, and Weingast (1990) and Greif, Milgrom, and Weingast (1995), argue that late medieval institutional forms (the *lex mercatoria* and merchant guild, respectively) came into being for the purpose of promoting trust and alleviating defection problems among merchants and traders.

All of this work argues, more or less explicitly, that institutions come into being *to serve the function* of producing trust, cooperation, or both. In so doing, it sometimes does violence to the empirical record. Terry Moe (2005), points out, for example, that *contra* North and Weingast, the Glorious Revolution was not under any reasonable interpretation motivated by the desire of the English Crown to make a credible commitment to its debtors. Instead, "democratic institutions had to be *forced*

[4] However, he specifically excludes trust from his account (Williamson 1993), arguing that economic decisions turn on calculativeness (in contrast, he argues that personal relations involve more-or-less unconsidered trust). For counterarguments, see Hardin 2002.

[5] Cf. Schelling's (1960) discussion of the value of the right to be sued.

on the monarchy."[6] In order to discover when functionalist accounts of institutions provide a good explanation, it is important not to confuse institutions' effects with their causes.[7] Sometimes institutions are indeed created in order to underpin cooperation. But if one wants to demonstrate this, it is not sufficient simply to demonstrate the effect (supporting trust or cooperation). It is also necessary to show that the effect was what motivated those who chose the institution (in situations of direct and deliberate choice) or was what selected the institution for survival over other possible institutions (in situations where institutions emerge indirectly from individuals' aggregated choices). Furthermore, in most social dilemmas, cooperation might be underpinned by a variety of different institutions (Hall 2005), which may differ, say, with regard to how the benefits of cooperation are distributed. The main prediction of cooperation-oriented functionalist accounts – that the institution chosen will be the one that maximizes the fruits of cooperation – is manifestly falsified in many historical circumstances. Institutions that are less "efficient" in terms of how they underpin cooperation but that provide distributional benefits to powerful actors are commonly selected over more efficient institutions (Knight 1992). Finally, exactly because these theories ground institutional creation and change in the need to maximize the benefits of cooperation, they make it difficult to create a proper theory of the consequences of institutional variation for forms of cooperation, suggesting that the lines of causality run from cooperation and cooperative problems to institutions, when there is good theoretical reason to believe that the causal arrow points in the other direction (Knight 1995).

These and other problems have led many historical institutionalists to reject explanations of institutions that see them as "structures of voluntary cooperation" (Thelen 2004, 25) in favor of explanations that provide more scope for contingency and particularity.[8]

[6] Moe 2005, 223, emphasis in original.

[7] Although Oliver Williamson (2000, 611) claims that "efficiency arguments have mainly prevailed over power interpretations because the latter are tautological," the former set of arguments are perhaps rather more prone to tautology than the latter.

[8] Sometimes, historical institutionalists have used the prevalence of functionalist explanations within rational choice to make sweeping indictments of rational choice theory as a whole. To the extent that there is a substantial body of rational choice theory that is explicitly nonfunctionalist, this is unwarranted. See further the debate between Paul Pierson (2000) and Gary Miller (2000).

Thus, for example, Paul Pierson (2004) points to the limitations of "actor-centered functionalism," stressing the gap between the intent of those who designed an institution and the various effects that the institution may have. Kathleen Thelen (2004) shows that the intentions of those who created the German vocational training system tell us very little about its later consequences and historical evolution.

Many authors in the historical institutionalist tradition have embraced path dependence as an alternative explanatory framework to rational choice explanations (Mahoney 2000; Pierson 2004). Path dependence, especially in its original formulation (Arthur 1994), provides a theory of how early choices may have important long-term consequences. Institutions can be "sticky," even in circumstances where they have clearly outlived or outgrown the intentions of their original creators (Gamm and Shepsle 1989).

However, scholars of historical institutionalism continue to face their own theoretical challenges. On the one hand, by seeking explicitly to show how institutions' development and consequences are distinct from the intentions of their founders, they forego cooperation-oriented functionalism in favor of an account of institutional change that lays a heavy emphasis on contingency and unforeseen circumstance. Path dependence theory still borrows heavily from Brian Arthur's (1994) mathematical treatment of Polya urn processes, in which chance events (the selection of balls of a particular color in Arthur's analogy) early on have a very substantial effect on the path selected over the longer term (Pierson 2004). On the other hand, once a particular path has been chosen, path dependence has difficulties in describing the circumstances under which this path might be deviated from in the future. As a result, many path-dependence accounts are overly deterministic, discounting the possibility that change may occur once a path has been found (Crouch and Farrell 2004). Even the most sophisticated path-dependence accounts (e.g., Pierson 2004) continue to be rather better at explaining why institutions persist than in explaining how they are selected in the first place or how they change in response to shifts in circumstances (Thelen 2002; 2004). This tendency is exacerbated by the frequent tendency of path-dependence accounts to invoke the broad concept of path dependence as a sufficient explanation in itself rather than detailing the specific mechanisms through which institutional reproduction and change takes place (Thelen 1999; 2002; for a valuable recent exercise in identifying such mechanisms, see Hacker 2005). In general, the approach is still better at providing actual accounts of how change occurred than predictions of which changes in outcomes are likely given specific changes in underlying conditions. While path-dependence and related approaches offer an exciting research agenda, they do not currently provide a sufficiently complete account of the sources of institutional change for predictions.

An alternative approach offers some of the advantages of path-dependence theory (viz., an analytic means to distinguish institutional origins from the socially beneficial effects that they may or may not have) together with a more particular focus on the circumstances under which they come into being and under which they change. Power-based accounts of institutional change focus on (1) the distributional interests of actors and (2) the differential ability of actors to act upon those interests. Much of the literature in historical institutionalism adopts this approach. For example, Charles Tilly's (1985) classic account of the origins of the state focuses on elites' desire to extract revenues from the populace and their ability to do so through the threat of violence. More recently, a vein of rational choice scholarship has adopted a variant of this perspective, examining both macroprocesses of state building and revenue extraction (North 1981; 1990; Levi 1988) and microlevel instances of coercion (Moe 1987; Knight 1992; Levi 1997; Brehm and Gates 1997). Much of the more microfocused work in this vein builds upon basic results from mixed-motivation bargaining games (the so-called "battle of the sexes" game).[9] Power in these contexts becomes bargaining power – the ability to drive a deal that favors your interests as much as possible, if necessary to the detriment of other actors.[10] Thus, one may

[9] See, for example, Krasner 1991.

[10] Bargaining power will in turn be driven by the "breakdown values," which is to say the payoffs that actors would receive if they were not able to agree on a mutually acceptable coordination point. Actors having few attractive alternative options in the case of a breakdown are likely to be in a weak bargaining position; they will be prepared to make substantial concessions in order to avoid this. Actors having many attractive alternative options in the case of a breakdown are likely to be in a

expect that under most circumstances, actors will coordinate on bargaining solutions that favor the interests of more powerful actors, if necessary at the expense of less powerful actors.[11]

The bargaining approach argues that these power asymmetries and inequitable solutions to bargaining games are likely to be instantiated in the form of institutions over the longer term. This is most obviously true in the case of formal institutions (Krasner 1991), where more powerful actors can directly press in negotiations for concessions that favor their particular interests. However, it is also true of informal institutions too, where institutions emerge from interactions among actors over time instead of from direct negotiations. Here, as actors begin to form stable expectations about the consequences of bargaining, the power asymmetries between actors will become instantiated over time in informal institutional rules, which will dictate how one set of actors should behave with respect to another.[12] For example, Jean Ensminger and Jack Knight (1997) examine the emergence of the informal norm of bridewealth among Orma cattle herders in Kenya. They argue that this norm reflects the bargaining power and interests of wealthy older men, who have traditionally controlled most of the resources in this society and who use it to ensure a plentiful supply of wives for themselves and their sons. If women (who are in a weak bargaining position) could act on their preferences unconstrained by this norm, they would be more likely to choose mates closer to their own age – and less likely to choose polygynous unions.

This account furthermore provides a straightforward theory of institutional change, predicting that institutions are likely to change under two conditions (Knight 1992). First, we may expect to see

much stronger bargaining position. They will thus be able to press for coordination points that reflect their preferences more than those of weaker actors.

[11] This argument is borrowed from Jack Knight (1992; 1995); see also Farrell and Knight 2003; unpublished. Knight (1995) uses a mixed-motive coordination game to explore the difference between convention accounts (borrowing from Schelling's [1960] work on salience), contract accounts (in which actors seek to create contracts to solve Prisoner's Dilemma-type problems), and bargaining accounts. Knight argues that if one believes actors to be genuinely self-interested, convention and contract accounts are only likely to have explanatory weight under relatively stringent conditions. Bargaining accounts will have more general application.

[12] For an extended account, see Knight 1992.

institutions change when there is change in actors' bargaining power. Actors who were previously weak but who become stronger (acquire more attractive alternative options in the case of breakdown) may be expected to press successfully for changes in institutions so as to improve their share of the overall distribution of benefits from cooperation. Actors whose bargaining power is weakened (so that they have fewer or less attractive alternatives in the case of breakdown) will have little choice but to make concessions. Again, Ensminger and Knight (1997) offer an empirical illustration. As bargaining power has shifted from wealthy older males into the hands of younger males, the practice of bridewealth has apparently begun to decline, just as the bargaining perspective would predict.

Second, even when bargaining power is held constant, we may expect change if there is change in the underlying preferences of actors. If powerful actors decide that they prefer a different cooperative outcome to the one they preferred previously, they will likely be in a good position to press other actors to change their strategies too, so that all actors converge on a new cooperative outcome that better reflects the changed desires of the more powerful actors.

The bargaining approach has some important weaknesses. Like functionalist accounts, the bargaining approach links institutional emergence to cooperation, in the very broad sense that institutions would not exist if it were not for the possibility that humans might gain from cooperating together in a given setting. Thus, it implies that actors are bargaining over a mutually beneficial outcome and that the stakes are the distribution of the benefits from cooperation. Some exercises of power (viz., the use of the threat of physical harm to persuade me to give up material resources) do not involve cooperation except in the most trivial sense of that word (Moe 2005) and are poorly explained by this theory (they may be better accommodated by historical institutionalist cognate theories, such as Tilly's). Further, it arguably overestimates the extent to which institutions are the product of conscious decision rather than iterative adaptation to changing circumstances and (in many cases) actors' failure to consider the long term outcomes of their institutional choices.[13] Finally, in its current form, it sometimes pays short shrift to

[13] Although for a bargaining account that stresses unforeseen consequences, see Johnson 2005.

the microprocesses through which power relations become translated into institutional change (a point I return to in Chapter 5).

Despite these limitations, the bargaining account provides a more supple account of where institutions come from and how they are likely to change than either functionalist or path-dependence approaches. On the one hand, it avoids the former's tendency toward intellectually muddy and post hoc explanations, in which the fact that an institution has (perhaps unanticipated and unintended) socially beneficial consequences is treated as a sufficient explanation of why this institution came into being in the first place. On the other hand, it offers clearer and less ambiguously testable predictions than path-dependence accounts, which typically stress the role of contingency and unanticipated consequences in determining paths of institutional development.[14] By focusing on actors' individual interests, how they may seek to use their bargaining power to achieve those interests, and how this affects institutional creation and change over time, we can make arguments about how institutions affect cooperation (and, as I discuss below, trust) without either starting from the premise that institutions come into being in order to underpin cooperation or having to forgo the possibility of making testable empirical predictions.

Not only this, but we can begin to properly examine how power relations (as mediated through institutions) affect the forms of trust and cooperation that are found in the political economy and elsewhere. Because much of the literature on trust and cooperation is directly or indirectly functionalist, it has tended to underestimate the extent to which power relations play an important role in determining who can trust whom.[15] Power relations may indeed exert an important influence on relations of trust and cooperation, including those power relationships that are institutionalized. I develop this argument at greater length in Chapter 5.

[14] Contingency plays a less prominent role in more recent applications of path dependence theory such as Pierson (2004). However, much work remains to be done on elaborating the mechanisms through which paths are set and sometimes abandoned. See Crouch and Farrell 2004 for further discussion.

[15] I note that Robert Putnam's work on social capital (1993) is a partial exception despite its functionalism; his discussion of how "horizontal" and "vertical" relations structure cooperation provides at least the beginnings of an account of how power enters into relations of cooperation.

2.4 INSTITUTIONAL CONSEQUENCES AND EXPECTATIONS

If our accounts of institutions' development have problems, our accounts of the consequences of institutions for trust and cooperation are equally vexed. This is especially true of rational choice accounts. The development of rational choice institutionalism has gone hand in hand with advances in the application of game theory to the understanding of social phenomena. Thus, one can identify two main rational choice approaches to institutions, each of which seeks to make arguments about their effects by treating them as features of games.[16] Structural approaches (which have their ancestry in the concept of "structure-induced equilibrium") treat institutions as part of the game form; here, institutions have consequences because they dictate that actors converge on certain equilibria rather than others. "Institutional equilibria approaches" take the opposite tack, treating institutions as social equilibria that have emerged over repeated iterations of relatively simple games. Here, institutions have effects for the same reason as other equilibria – they are stable (as long as their underlying parameter conditions are stable) because no one wishes to deviate from them, given what everyone else is doing. The vast majority of rational choice arguments about institution-induced cooperation employs one or the other of these approaches; as I will argue below, neither provides a very good account of how institutions can induce trust.

The dominant approach to the understanding of formal political institutions treats them as part of the game form.[17] Thus, institutions have effect through structuring the kind and sequence of choices available to actors, in a way that might lead them to select determinate equilibria. Structural approaches have laid heavy emphasis on exploring the circumstances under which particular institutions (modeled

[16] This distinction owes a clear intellectual debt to Randall Calvert's (1995a) overview of institutional theory, but it complicates his twofold distinction between institutions as parts of the game structure that induce equilibria and institutions as equilibria. The idea that institutions as external mechanisms of equilibrium selection, outside the game form, is quite different from the first two, and provides a more supple means of capturing the effects of institutions on tricky forms of expectation such as trust.

[17] This approach stemmed from the dissatisfaction of scholars such as Kenneth Shepsle and Barry R. Weingast with social choice theory's lack of attention to the ways in which social context shaped outcomes, and consequent claims that "chaos" and cycling should be pervasive in politics (Shepsle 1989).

as rules of procedural choice) may create stability over time, avoiding the cycling problem. Much attention has been paid to examining how institutional rules lead to equilibria in formal legislative settings such as the U.S. Congress and the European Union's codecision process (Shepsle and Weingast 1981; Tsebelis and Garrett 2001).

Many scholars have sought to begin from the other end, analyzing institutions as equilibria within games. This approach had its beginnings in the work of Andrew Schotter (1981). Prominent recent applications include Milgrom, North, and Weingast (1990), and Greif (1994); perhaps the most sophisticated discussion of its implications can be found in the work of Randall Calvert (1995a; 1995b; 1995c). Unlike the structure-induced equilibrium account, work in this vein usually examines institutions in the context of infinitely repeated games, examining whether a set of strategies corresponding to the institution in question can be sustained as an equilibrium under various parameter values. As is well known, many games have more than one equilibrium, limiting game theory's ability to predict outcomes. This problem is especially vexing in the case of infinitely repeated games, where folk-theorem results under a wide variety of parameter values "state, in effect, that nearly any reasonable outcome can be sustained by some equilibrium of individual strategies."[18] Institution-as-equilibrium approaches seek to turn the folk theorem on its head, demonstrating that a particular combination of strategies making up an institution can be supported as an equilibrium under a given set of parameter values, often without inquiring in any systematic fashion into why this equilibrium, rather than another, was selected.

This approach treats institutions' effects as similar to those of other equilibria. That is, once actors have reached a particular equilibrium through a particular combination of strategies, no individual actor has an incentive to deviate from that strategy given that other actors are playing the strategies that they are playing. Under the institution-as-equilibrium approach, institutions will be self-enforcing and more-or-less static patterns of interaction. It is thus notably difficult to disentangle the effects of the institution from the institution itself: the two are bound up in each other so that the institution is its own effect. One recent variant of this approach has modified this definition in

[18] Ordeshook 1990, 29. See Davies 2006 for a significant corollary.

important and fruitful ways, providing arguments about how institutional equilibria may prove self-reinforcing or self-undermining over time, depending on how they affect "quasi-parameters" (Greif and Laitin 2004). However, it still continues to adhere to a somewhat confusing definition of the term institution, in which it is not entirely easy to discern where the institution ends and its effects on social behavior begin.

Both of these arguments about institutional effects have considerable difficulty in incorporating trust. Under the structure-induced-equilibrium approach, institutions may indeed induce actors to cooperate, but it is difficult to describe this cooperation as being rooted in trust. Instead, it is a more or less mechanical consequence of the institutional rules in a given setting. Here, cooperation is mere institutional compliance. The institution-as-equilibrium approach argues that institutions are self-enforcing equilibria, from which no actor has any incentive to deviate, given the strategies of other actors. As David Kreps (1990a) and Gary Miller (1992; 2001; 2004) have argued, equilibria in infinitely repeated games provide us with a very useful way of understanding how informal trust and gift exchange can be supported by rational actors. But in a very real sense, the institution-as-equilibrium approach provides us with a theory of effects without causes, or, even worse, it fails to distinguish *between* the effects and the causes that we might reasonably wish to explore. If our theory of how institutions induce trust effectively reduces trust to a set of institutional relations that can be modeled as an equilibrium in some underlying game, then speaking about the relationship between institutions and trust is tautologous – institutions and trust are one and the same thing. Thus, while studying equilibria in repeated games can teach us valuable lessons about trust, we need to supplement this body of theory with another that (1) distinguishes between institutions and their outcomes and (2) better explains *which* equilibria are chosen under which circumstances.

2.5 INSTITUTIONS AND TRUST

I have set out two quite different and apparently unrelated literatures – the literature on trust and cooperation and the rational choice literature on institutions. Neither of these literatures seems well suited to providing a proper account of the relationship between institutions and trust on its own. But given their very different starting assumptions, how may they be reconciled with each other?

On the one hand, most theories of institutions appear to provide a rather deterministic account of the sources of cooperation. This is trivially true in the structure-induced equilibrium account, in which institutions are determinate structures within a game that conduct toward equilibrium outcomes. But it is also true of the institutions-as-equilibria account; although scholars such as Paul Milgrom, Douglass North, and Barry Weingast claim to provide an account of the sources of the trust necessary for efficient exchange, they are not really talking about trust but instead about a very narrowly defined set of mutually reinforcing expectations.

But on the other hand, our theories of trust among individuals do not provide us with enough traction to understand how institutions may directly affect trust relationships. Hardin's account provides a theory of trust that is rich enough to incorporate many of the factors that influence trust and trustworthiness in real life, including nonrational motivations (Hardin 2002). Yet because it grounds trust and trustworthiness in personal relationships, it has considerable difficulty in incorporating the effects of institutions. Some institutional effects, most notably institutions' capacity to support the dissemination of information (Levi 1998), may very clearly contribute to alleviating distrust (although under Hardin's somewhat different account, their positive effect in increasing trust is more ambiguous). But others may not. In particular, social relationships such as those between state officials and citizens might seem to possess the characteristics of trust as the encapsulated interest account describes it (Levi 1998), but they nonetheless do not typically fall under Hardin's more stringent definition. They do not involve personalized relationships and the knowledge that such relationships conveys.

Thus, in order to develop a proper account of the relationship between institutions and trust, it is necessary to address two key problems. First, one must provide a theory of institutions that incorporates some degree of ambiguity, as conventional game theoretic accounts do not. As both Gambetta (1988) and Lorenz (1999) argue, when institutions have entirely determinate effects on expectations, they do not, by virtue of this fact, create trust. Determinate accounts in which institutions necessarily and ineluctably lead to determinate equilibrium outcomes are thus an inappropriate basis for a theory of institutions and trust.[19] Second, while doing this, one must provide a theory of trust that does not rest exclusively on personal relations and can thus encompass the less personalized (but still somewhat specific) expectations that institutions give rise to.

We can begin to explore institutional effects in a less deterministic fashion if we consider how they may work as *external* mechanisms of equilibrium selection.[20] Within game theory, various scholars have sought to tackle the problem of equilibrium selection through increasingly exotic refinements of the concept of equilibrium. However, these internal approaches have at best succeeded in somewhat reducing the degree of uncertainty, and, perhaps, in the case of some of the more abstruse refinements, not even that (Kreps 1990b). Another approach, building on the contributions of Thomas Schelling, has been to identify mechanisms external to game theory that single out some equilibria and make them more "salient" (Schelling 1960) to players than others, and thus more plausible outcomes of social interaction. Some equilibria may have "focal point" qualities; that is, they stand out in some way from other possible equilibria. Provided that this is common knowledge among all actors, one may expect them to converge, *ceteris paribus*, on the equilibrium with these qualities rather than another.[21]

Institutions provide just such a mechanism, helping actors to decide which of a number of possible equilibria they should converge upon.[22] Under this argument, institutions are neither part of the game form nor an equilibrium within a game. Instead, they are phenomena external to the game that have important consequences for the strategies

[19] As discussed below, the institution as equilibrium selection approach does not require this degree of certainty.

[20] Here, I seek to render more explicit a line of argument that has previously been developed by Jack Knight (1992; 1995). For further discussion, see Farrell and Knight (unpublished).

[21] See Sugden 1995, Calvert and Johnson 1999, and Schiemann 2000 for more formal discussions of focal points and salience.

[22] The literature also identifies a number of other mechanisms, including culture (Kreps 1990a, Johnson 2002), leadership (Miller 1992), and judicial precedent (Calvert and Johnson 1999).

chosen within games, because they help actors to converge on one equilibrium rather than another in games where there are many possible equilibria on which actors might converge. Treating institutions as mechanisms of equilibrium selection allows one (1) to identify the relevant institutions in a given context and (2) to show how these institutions help actors to converge on one mutually reinforcing pattern of behavior rather than another. However, it also requires one to identify the specific mechanisms through which institutions make some equilibria more salient than others.

Why might institutions have this effect? If institutions are indeed sets of rules, then they may provide two important kinds of information. First, they will provide information to people about how general classes of actors are supposed to behave in specific situations.[23] Not only do formal and informal institutions give actors information on how they are supposed to behave in a given social situation but they also give actors information on how they may expect others (belonging to the same or other general classes) to behave too. Most pertinently, they may give actors information on how others are likely to react (and perhaps punish them) if they fail to live up to the social expectations that are embedded in a particular situation. Thus, under this argument, institutions are reinforced not by a general sense of "appropriateness" (March and Olsen 1989), but by an awareness of the consequences should there be deviation from the behaviors prescribed by the institution. Second, certain kinds of institutions – that is, institutions governing the ways in which actors communicate with each other – will affect the ways in which information about specific others will be distributed throughout a given community and society, and how this information is interpreted. Institutions that support the widespread dissemination of pertinent information about others' behavior and intentions will make it easier for individuals to trust other individuals in a grounded way (when those individuals' past behavior and apparent intentions suggests that they are likely to be trustworthy with respect to the individuals and matters at issue), and to identify potentially untrustworthy individuals (where those individuals' past behavior and apparent intentions carry the opposite

[23] My argument about institutions in this chapter draws in part on Farrell and Knight (unpublished).

connotation). Again, this helps conduct actors toward equilibria in uncertain situations. I deal with each of these mechanisms in turn.

An institution, if it is to have reasonably predictable social consequences, should be common knowledge among all the relevant actors. All actors should have a good idea not only of how they are supposed to behave, given the existence of the institution, but also of how others are supposed to behave too. Given the above, institutions can help actors to converge on an equilibrium from which no actor has any incentive to deviate, provided that all other actors behave in the expected way. If the institutions are appropriate, then this equilibrium may involve high-trust patterns of gift exchange and reciprocity (Miller 1992).[24] If the institutions are not appropriate, we may expect equilibria in which actors do not trust each other and perhaps actively distrust each other. Thus, the argument that institutions serve as a means of equilibrium selection provides a way to distinguish the institution itself (the rule or rules that allow individuals to interpret how others are likely to behave) from the patterns of trust or cooperation that the institution may conduct toward.

Some important implications follow from the claim that institutions are best considered as external mechanisms of equilibrium selection. First, as the distributional theories of institutional emergence described earlier imply, we have no reason to expect that social institutions need be either equitable or economically efficient. The informal rule of the pre–civil rights South that a black person would leave the sidewalk to make way for a white person coming in the other direction qualifies as an institution under this definition, as do the formal and informal institutions that obliged black people to use separate public facilities from white people. So too do the informal expectations governing bribery qualify as social institutions; to the extent that everyone in a particular city or country knows that a traffic policeman who pulls me over is expecting a bribe, this is an informal rule that will structure behavior.

Second, exactly because institutions are social rules, they not only are common knowledge among the relevant group of individuals, but they are typically specified with regard to *classes* of actors rather

[24] I discuss the circumstances under which institutions are likely to be appropriate or inappropriate at greater length in Chapters 4 and 6.

than to individuals.[25] My personal precepts as to how best to interact with a specific colleague in my department are not an institution. However, the formal rules governing interactions among the members of the department most certainly are. So too are the informal rules governing social intercourse within the department; whether or not, for example, professors leave their doors open or closed. Typically, an institution will take something like the following form: when an actor (or actors) of type *A* is in a social situation of type *C*, together with an actor (or actors) of type *B*, then, the appropriate action for an actor of type *A* to take is *X*, and the appropriate action for an actor of type *B* to take is *Y* (here, I use uppercase letters to denote classes of actors, situations, and actions, as opposed to specific actors, situations, and actions, which I continue to denote with lowercase letters). If I myself am an actor of type *A* in this situation, then I will know that I am expected to do *X*; I will also expect another individual of type *B* to do *Y*.[26]

Third, as an implication of the above, there is frequently some scope for ambiguity in the effects of institutions. When an institution has material social consequences, it is because people reason from the abstract rule to the concrete situation and set of actors that they are confronted with, and take action accordingly. As H. L. A. Hart notes of the law:

> its successful operation over vast areas of social life depends on a widely diffused capacity to recognize particular acts, things, and circumstances as instances of the general classifications which the law makes. (Hart 1994, 124)

However, precisely because the rule is abstract, it may not be clear exactly how it applies to a given situation. It may be that the material situation is not precisely covered by a given institutional rule (Hart 1994), because the categories described in the relevant rule do not seem to map very well onto the particular issue at hand. It may be

[25] Here, I strengthen H. L. A. Hart's (1994, 124) claim that "the law must predominantly, but by no means exclusively, refer to *classes* of person, and to *classes* of acts, things and circumstances" and extend it to include institutions other than legal institutions.

[26] I lay out one of the simplest forms that such institutions might take here; obviously the same basic arguments will apply to institutions in which there are more than two types of actor involved, or in which there is only one type of actor (e.g., Type A actors dealing with other Type A actors).

that there are a number of institutional rules, instantiating conflicting expectations, which might possibly be said to apply in a given situation, so that it is difficult to decide whether one rule or another rule should be followed. It may be that it is difficult to identify which class another actor belongs to and thus to predict that actor's behavior (and modify one's own expectations accordingly).

In this book, I will bracket many of these sources of uncertainty in applying rules to concrete social situations; they are not highly relevant to the causal relationships that I am interested in. However, I will note that there are likely to be systematic differences in the degree of ambiguity that characterize different *kinds* of institutions. The existing literature (Ostrom 1990; North 1990; Knight 1992) argues that there is an important distinction between formal institutions and informal ones. The former are written rules – laws, contracts, and the like, which are typically backed up by the enforcement capacity of the state and by systems of adjudication in which specialized actors (such as judges) resolve conflicts of institutional interpretation. The latter are typically unwritten rules – informal social norms – which are not backed up by the enforcement power of the state but by informal sanctions delivered by other community members.

While considerable attention has been paid to differences in enforcement mechanisms between informal and formal institutions (Gibbons and Rutten 2004; Dixit 2004) there has been very little discussion of another crucial difference between the two: their levels of ambiguity. Formal institutions, which are written, tend to have a high degree of certainty attached to them (Carey 2000). Those who are subject to them will have a good idea of how the institutions will apply in specific instances, who will enforce them, and how. By the same token, they are usually relatively narrow in scope. Informal institutions, in contrast, tend to provide less certainty in application and interpretation, but they are also broader in range, and more easily adopted to unforeseen contingencies. Thus, in short, formal institutions tend to be considerably less ambiguous in their scope and extent than informal ones. I develop the implications of this difference for trust and cooperation in Section 2.6.

This account of institutions is rather more easily reconciled than dominant rational choice accounts with a rich account of trust along the lines that Hardin offers. Even so, if we wish to adopt Hardin's arguments about trust to situations in which expectations derive from institutions rather than personal relationships, we must first examine more precisely the particular kinds of information that institutions provide. This information comes in two distinct forms: (1) knowledge instantiated within the institutions themselves and (2) reputational information conveyed by actors to other actors under the rules of specific institutions.

First of all, institutions instantiate commonly held beliefs about how others are likely to behave in given social situations. Thus, as already noted, the information they provide in and of themselves is of a quite specific sort – information on how classes or *types* of actors are likely to behave.[27] This information is quite distinct from the information about more intimate relational incentives that Hardin and others focus on. Institutions provide information about the range of types to be found in a given society, and how they will interact with each other, in part because they instantiate beliefs about the ways in which different actors are likely to behave.[28] In the absence of intimate acquaintance, and the forms of knowledge that flow from it, the beliefs instantiated within institutions provide crucial information about how different types of actors are likely to behave across a variety of situations. Even where actors are dealing with others who are personally well known to them (or whose reputation they are aware of; see below), institutions may provide important information – the type (i.e., social category) of actor that I am dealing with may be as important a predictor of his or her actions and reactions as his or her specific personal attributes.

Consider, for example, the following mechanism. An actor *a* is aware of the existence of social rules applying to a specific type of actor, *B*, which suggest that type *B* actors will be trustworthy in interactions

[27] The information that institutions instantiate is obviously different from the information that some institutions (e.g., institutions covering communication among actors) *convey*. In other words, institutions may embody important social information in and of themselves. Certain kinds of institutions, which shape the ways in which actors do or do not communicate with others in their community, will also have consequences for the ways in which information (for example, information about the behavior of specific others), is transmitted or not transmitted. Both of these kinds of information can affect trust – but in quite different ways. I turn to this in the next section.

[28] My argument here borrows from Johnson (2002).

with *a* (or more broadly, actors of type *A*, which includes *a*) in situation *C*. It is reasonable to expect that *a* will trust actors of type *B* in situations that resemble *C*; she will expect that when she encounters actors of type *B* in such situations, they will take her interests into account. For example, in the United States, people of certain social categories (upper-middle and upper-class members of dominant ethnic groups) may have good reason to believe that members of the police will behave in a trustworthy fashion in a wide variety of interactions. These beliefs are intimately associated with informal yet pervasive institutions that dictate how members of the police interact with different types of citizens (for example, in the United States, well-dressed Caucasians driving expensive cars are usually not stopped for questioning without good cause).[29] Members of this social category may reasonably trust the police in many situations, reasoning on the basis of existing institutions that they belong to a type that the police has good reason to treat in a trustworthy fashion. More abstractly, in such situations *a* knows that she is a member of the class *A*, and is dealing with a member of type *B*; the social beliefs instantiated in institutions give her good reason to anticipate that members of type *B* will behave in a trustworthy fashion toward members of type *A* across a variety of situations. Here, the logic is similar to Hardin's personalized trust but on a broader scale of social abstraction. For Hardin, *a* trusts *b*, with regard to matter *c*, because she knows that *b*'s interests encapsulate her own in this matter. Under the mechanism that I describe here, *a* trusts *b* with regard to situations resembling *C*, because *a* is a member of type *A*, while *b* is a member of type *B*, and *a* knows from existing institutions that type *B* actors have good reason to be trustworthy when interacting with type *A* actors in situations that resemble *C*.[30]

[29] For empirical evidence from the United States, see Weitzer and Tuch 1999.

[30] Of course, mechanisms of this sort may lead to distrust rather than trust. If *x* knows that she is dealing with an actor of type *B**, and social institutions instantiate beliefs regarding type *B**'s untrustworthiness across various matters, she is unlikely to trust the actor with respect to these matters. Such beliefs can be self-reinforcing. If actors perceive that they belong to a specific type that is regarded as untrustworthy, and consequently are unlikely to enjoy the benefits of social cooperation, they may quite reasonably decide to behave in an untrustworthy fashion on those rare occasions when others trust them, since the longer term benefits of a

Under this argument, institutions may provide information that is *useful* to actors trying to determine whether to trust each other, without necessarily *determining* their actions in ways that would drive out trust. Here, Hardin's vocabulary may be adapted to provide a clearer understanding of the issues involved – and of the difference between institution-induced expectations and trust. First, consider the case where *a* and *b* find themselves in a situation *c*, which is explicitly covered by an existing institutional rule setting out how agents of type *A* and *B* should behave under circumstances *C*. Under this set of conditions, it is difficult to describe *a*'s expectations regarding *b*, or *b*'s expectations regarding *a*, as involving trust. Their mutual expectations are directly induced by the institution in question. Thus, when institutional rules apply directly and unambiguously, we should speak of institution-induced expectations and of institutional compliance, not of trust.[31]

However, consider an alternative case in which *a* and *b* find themselves in a situation *c**, which resembles *C* in important respects, but is not precisely and unambiguously covered by it (as already noted, such situations are endemic; the application of abstract institutional rules to concrete situations is rife with ambiguity). In such contexts, the relevant institutional rule does not compel or induce compliance directly, as it is not entirely clear whether the rule should or should not apply under the given circumstances. Here, institutions may still play a very important role in affecting actors' expectations and behavior but not a determinative one. Existing institutions may provide actors after the fact with a means of interpreting situations that were not specifically anticipated, but that in some important way resemble the situations covered by existing institutional rules (Calvert and Johnson 1999). Rather than inducing behavior directly, these institutions work in the same way as the corporate cultures described by David Kreps (1990a) and Gary Miller (1992); they provide a focal point on which actors' expectations regarding appropriate behavior may converge. If actors *a* and *b* find themselves in situation *c**, and they are uncertain

reputation for trustworthiness are unlikely to accrue to them. See, more generally, Knight 2000.

31 For further development of the distinction between compliance and trust, see Farrell and Knight (unpublished).

of how to behave, we may reasonably expect that they will draw upon their knowledge of how type *A* actors and type *B* actors interact in *C* in determining the appropriate actions. Where the situations are analogous, actor *a* will be more likely to behave in accordance with the behavior prescribed for *A* in the relevant institutional rule, and actor *b* will be more likely to act in accordance with the behavior ascribed to *B*. If the institution in question prescribes that actors of type *B* behave in a trustworthy fashion toward actors of type *A* (or vice versa), actor *a* will have some grounded reasons for trusting *b* in situations that are not precisely C but are analogous to C. While the rule does not determine actors' behavior, it does give them initial guidelines about how to behave in unexpected circumstances. To the extent that there are few other reliable sources of information regarding types of actors and how they will behave, the beliefs instantiated in institutions will be compelling focal points – in a situation of uncertainty, they allow actors to interpret situations, to predict how others are likely to behave, and to act accordingly.

We may also expect that ambiguity over whether an institution covers the situation in question will lead to ambiguity over whether or not the institution will or should be enforced under unanticipated circumstances. Thus, we may expect that the likelihood of enforcement will vary with the degree of similarity between c* and C. This will have the effect of making the relevant institutional rule more attractive as a focal point – if actors reasonably fear that others may enforce the institutional rule as if c* fell under C, they will be more likely to behave as if c* indeed falls under C. There is relatively little role for trust in circumstances where c* is nearly identical to C, and the institutional rules and enforcement mechanisms applying to C are almost certain to apply. However, there will be very many possible situations in which c* is only somewhat analogous to C, and which will give rise to some uncertainty as to whether or not the enforcement mechanism will be invoked. Under these circumstances, we may expect that the possibility of enforcement will increase the focal power of the institution, without at the same time involving that certainty of application implied in the notions of confidence and of institution-induced compliance.

Furthermore, we may expect that institutions will not only affect trust through the general information that they instantiate (over how different classes of actors may be expected to interact with each other) but also through the more specific information that they allow actors to convey to each other about *particular* actors' past behavior and intentions. Much of the game theoretic work on institutions (Milgrom, North, and Weingast 1990; Greif 1994; Greif, Milgrom, and Weingast 1995; Calvert 1995a; 1995b; 1995c) seeks to model institutions that oblige actors both (1) to communicate information to others about the honesty or dishonesty of other actors whom they deal with and (2) to carry out certain punishments on the basis of this information when actors default. Even if this literature is not really talking about trust so much as confidence, its insights into how institutions govern communication may be modified to help explain the sources of trust.

As game theorists point out, actors may communicate about their experience with specific others, and their perceptions of the honesty or dishonesty of those others in particular instances. The ways in which they communicate are likely to be shaped in important ways by existing institutions in two ways. The first is easy to accommodate within game theory – the consequences that institutions have for who communicates with whom about what. This is not to say that game theory is true to real life; game theorists' models of these institutions are far more rigid than the kinds of rules that we might expect to see in real life. For example, many game theoretic models envisage that when one actor is cheated by another, the first actor is compelled to report it, either to all actors within the community or to a central agent who can act as a clearing house for reports of bad behavior (see, e.g., Calvert 1995c). Real life institutions of communication are likely to be considerably less demanding, although they achieve similar results through less exigent means (such as the dissemination of gossip and the like). Even so, institutions do demonstrably shape the ways in which actors do or do not communicate to each other and the kinds of information that they convey when they communicate. Actors may be willing to communicate some kinds of information and not others or to talk to some actors and not others, depending on the relevant institutional rules governing who talks to whom, and under what circumstances.

The information that we and our cognates receive under relevant institutions may in turn affect our decisions over whether to trust or distrust specific others. This is because this information is likely to shape actors' *reputations* in important ways, through circulating information about their past behavior and apparent intentions throughout the relevant community. Reputation is not limited to the determinate kinds of information that game theorists model. It also provides a means for assessing the likely trustworthiness or lack of trustworthiness of specific others who are not personally known to you in situations that cannot entirely be anticipated. As David Kreps (1990a) notes in his discussion of corporate culture, firms often seek to build and maintain reputations so as to communicate to employees and others that the firm will be trustworthy in unexpected circumstances.

This suggests that institutions may have a second set of consequences for our trust in particular others that is rather more difficult to accommodate within game theory. Not only do institutions shape the ways in which we communicate with each other over perceived infractions, they shape our perceptions over what is an infraction (i.e., an action that might give its author a reputation for untrustworthiness) and what is not. Game theoretic accounts of institutions and cooperation do not, as a rule, problematize the question of what constitutes defection under the rules and what does not. Yet this is a recurrent problem for real life institutions, where the question of what is untrustworthy behavior and what is not, is not, as in standard game theoretical accounts, a given. Some forms of sharp practice are unlikely to harm an individual's reputation, even if they seem untrustworthy on the face; indeed, they may enhance it. Other infractions that might not appear any more harmful may incur serious reputational penalties. Here, institutions may serve to resolve ambiguity by providing a means through which individuals' actions may be interpreted as being trustworthy or untrustworthy.

Thus, institutions may lead to the sharing of information allowing us to make judgments about other actors' intentions, and in particular over whether these intentions are likely to lead them to take your interests into account under situations of a certain kind. Institutions, when they lead actors to share information over specific others' (or their own) past behavior and intentions, will thus affect these actors' reputations, and hence the willingness of others to trust them under various circumstances. They will also provide a lens through which actors can examine – and interpret – the acts of others as being trustworthy or untrustworthy.

Where the relevant institutions provide guidance regarding trustworthy behavior, and facilitate the easy spread of information, we may expect that actors in the community will have a good idea of the respective trustworthiness and untrustworthiness of particular others. We may also expect that actors will be more likely to behave trustworthily than they might otherwise, since they will reasonably fear that they will lose out on future opportunities of exchange if they behave in a trustworthy fashion. Where the relevant institutions do not facilitate the flow of information, we may expect that actors in the community will have much greater difficulty in determining which specific others are likely to be trustworthy or untrustworthy under particular circumstances. We may also expect that actors will have less grounded reason to behave in a trustworthy fashion themselves, since they will have less reason to fear that bad behavior will adversely affect their reputation and their ability to engage in future exchange.

Together, these two mechanisms provide us with an understanding of how institutions may influence expectations, but not in a determinate manner that would drive trust out. Trust has little independent explanatory power in situations where actors possess definitive and complete knowledge of how others will behave. In such situations, expectations are better characterized as confidence than as trust. Thus, institutions are likely to affect trustworthiness and thus trust, not in circumstances where institutions have direct and unambiguous consequences for behavior, but instead where there is some ambiguity in how or whether institutions should be applied.

First, they may do this because they instantiate useful information about the *general* classes of actors that we might encounter in a given society or community, from which we can draw inferences about how these classes of actors are likely to behave under circumstances that are not precisely described by the institution in question. In other words, they may affect trust by providing guidance over how actors should behave in situations that were not directly anticipated but that in some sense come to resemble those situations that were anticipated.[32]

[32] For a more formal discussion of how such social processes might work, see Calvert and Johnson 1999.

Second, they may conduct actors toward the sharing of information about the past behavior and apparent intentions of *specific* others. Institutions may both provide the means through which actors interpret others' actions and shape the ways in which they communicate with each other about those actions. Different institutional frameworks will lead to differences in what information actors share, and with whom. This in turn may have direct implications for trust – where institutions facilitate the easy spread of information, we may expect that it will be easier for actors to trust specific others who have acquired a reputation for trustworthiness in certain kinds of situation and to avoid others who appear to be less trustworthy. Where, to the contrary, institutions impede or block the flow of useful information about others' behavior and motivations, we may expect that it will be more difficult for individuals to place their trust in others and to distinguish trustworthy from untrustworthy individuals. We may also expect that individuals will have less incentive toward trustworthiness under these circumstances.

2.6 INSTITUTIONAL CHANGE, INSTITUTIONAL VARIATION, AND TRUST

The preceding sections of this chapter set out the building blocks of a theory of the relationship between institutions and trust, which, building on existing debates in institutional theory, sets out an account of institutional creation and change that distinguishes sharply between institutions' origins and their effects. It then argues that we can understand the effects of institutions best if we think of institutions as external mechanisms of equilibrium selection, working to make one equilibrium set of strategies or another appear more "salient" to actors in imperfectly anticipated situations, without necessarily dictating as such that they converge upon this equilibrium. Institutions may have this effect because they provide information about the different classes of actors in a society and how they are likely to behave under a variety of circumstances. This information allows people to make inferences about how others are likely to behave under circumstances that are not precisely covered by institutional rules but that are analogous to such situations. They can draw upon these inferences in order to decide whether they can trust others (who belong to broader social categories) to cooperate and behave trustworthily with regard to a particular matter.

What implications can be drawn from this set of propositions? While they seek to describe the relationship between institutions and trust in broad terms, they do suggest that we may expect variation in institutions will lead to variation in trust. This will be true with regard to institutional variation over time and space. The mechanisms set out above have specific implications for both kinds of variation.

First, we may expect that institutional variation over time will have important consequences for trust. As I have argued, the bargaining power approach provides considerable analytical leverage over the questions of where institutions come from and how they change over time. It argues, *inter alia*, that we may expect to see institutional change occur when there are changes in the relative bargaining strength of the relevant actors. Actors whose bargaining position becomes stronger will be better able to push for an increase in their share of the benefits of cooperation. Actors whose bargaining position becomes weaker will be less able to resist the demands of stronger actors and may see their share of the benefits of cooperation decrease. These shifts in bargaining power will be reflected in institutional change. Where institutions are negotiated formally, the more powerful actors will be able to press directly for concessions. Where institutions emerge from informal processes and repeated interactions within a community, they will be updated over time to reflect changes in relative bargaining power.

How will this affect trust? As noted, institutions provide information about different types of actors in a society and how they might be expected to behave. When institutions reflect substantial power asymmetries, they also provide important information about how more powerful actors and less powerful actors are likely to behave toward each other. Just as with Ensminger and Knight's Orma cattle herders, we can predict that as power relations change over the longer term, so too will institutions change and hence expectations over trust and cooperation. I develop these arguments in greater depth and apply them to the understanding of institutional change in Chapter 5.

Second, this set of arguments allows us to draw useful comparisons between different *kinds* of institutions, and how they are likely to affect trust and cooperation. Economists are beginning to use stylized accounts and game theory to explore the differences between economies in which formal institutions are the main force structuring economic interactions and economies in which informal institutions are the main force.[33] In the former, institutions are underpinned in the final analysis by the coercive capacity of a specialized third party, the state. Economic interactions are typically impersonal, in the sense that the specific identity of the transacting parties is strictly irrelevant to the transaction. What matters is not who the transactors are, or their past history, but that both transactors are subject to the coercive authority of the same third party. In the latter, institutions are not underpinned by a specialized third party so much as the collective coercive capacity of the community of actors in question. Actors who do not comply with the relevant institutions are likely to suffer bilateral or collective punishment.

This new literature has very considerable promise as a means of generating institutional typologies. However, it typically fails to consider the way in which both formal and informal systems not only differ in their enforcement mechanisms but also in their degree of ambiguity. As I have argued above, informal institutions are likely to be considerably more ambiguous in consequences and enforcement than formal institutions. The latter are written and are usually subject in the final analysis to the jurisdiction of specialized third party actors (such as courts of law), which usually rely on some version of precedent in order to interpret how institutions should be applied. *Ceteris paribus*, formal institutions will usually be relatively tightly drawn and clearly delimited in their consequences – actors will have a good idea beforehand of what the relevant institutions do or do not entail.[34]

[33] For an overview of this literature, see Dixit 2004. Important contributions include Greif 1994; 1999; 2005; Gibbons and Rutten 2004; and McMillan and Woodruff 2000.

[34] Three qualifications are in order. First, some formal texts, such as constitutions, are inherently ambiguous, insofar as they instantiate rights frequently in conflict with each other (Stone Sweet 2000). As a result they provide some considerable latitude to judges in constitutional courts to decide which right should prevail under which set of circumstances. However, this latitude is likely to become more tightly constrained over time, as judges' scope for creative interpretation is limited by one version or another of *stare decisis* and the binding (or at least indicative) force of precedent. Second, there will be some variation in the scope of ambiguity in different institutional systems. In contrast to Anglo-American jurisprudence, some continental European legal systems (Arrighetti, Bachmann, and Deakin 1997)

Informal institutions, in contrast, are unwritten and are not supported by specialized actors who are bound by strong norms of precedent and consistency. Instead, they are usually interpreted as they operate by the community as a whole. To borrow a term from Hart (1994), formal institutions usually have a tightly defined "core" of application. They provide clear information on exactly what actors of type A and type B are supposed to do across a relatively well defined set of situations, C. By the same token, they are likely to provide only limited information, if any, to individuals about what they are expected to do in situations c* that somewhat resemble C, but that do not fall under C as it is defined under the institution in question. Furthermore, in advanced industrial democracies, enforcement of these institutions will usually be limited only to those instances that clearly fall under C, or are very close to it; the specialized actors in question (courts of law and law enforcement) will be bound by their own norms and institutions.

Informal institutions are quite different. They are unlikely to have a tightly defined core of application – the extent of C is likely to be relatively diffuse. By the same token, they have a broad "penumbra"; they are more easily applied to situations c* that only somewhat resemble C (indeed, the boundaries between situations of type C and c* will be quite poorly defined). Thus they are likely to apply to a much wider (if more difficult to define) set of situations than formal institutions.

These differences have clear implications for trust. If, as I have argued, trust requires some degree of ambiguity, then we may reasonably expect that formal institutions will have far more limited consequences for trust than informal ones. While formal institutions may very clearly support cooperation on a very wide scale, they do so primarily through promoting confidence rather than trust. Their effects on expectations are far more tightly defined and more determinate

invoke a general duty of "good faith" in commercial transactions, which provides greater flexibility at the cost of an increased degree of ambiguity over how the duty would be interpreted in a legal dispute. Finally, the degree to which formal institutions provide certainty depends of course on the degree to which the state in question is prepared to enforce them consistently. This will vary with the rule of law, the extent to which law is applied systematically to those governing, and those who are governed. For an important treatment of ambiguity and arbitrary enforcement of formal institutions in English history, see Hay 1975.

than informal institutions. We may expect that under some circumstances formal institutions may be leveraged to make trust easier by removing potential sources of distrust. For example, contracts may be used to rule out certain kinds of exploitative behavior, making it easier for actors to trust each other in other matters that are not directly covered by the contract in question. However, these effects are indirect ones; they do not impinge on trust directly, so much as they make it possible by ruling certain kinds of behavior off limits.

In contrast, informal institutions may have more direct consequences for trust and cooperation. Their effects on cooperation are likely to be much more diffuse beforehand. However, in practice they provide guidance to actors, allowing them to converge upon equilibrium patterns of behavior. In this sense, they are very much like David Kreps's (1990a) notion of corporate culture or Edward Lorenz's (1988; 1992b; 1999) "moral contracts." By providing information to actors on how they and others are expected to behave, they provide a means of interpreting – and resolving – social dilemmas, even in situations that were unanticipated (but that bear some resemblance to situations that were indeed anticipated). They also provide a necessary interpretive framework for reputational effects by providing information on which kinds of behavior are regarded as "good" and which as "bad" (Kreps 1990a, Lorenz 1999). By the same token, however, much of their force comes from the expectation that these interpretations will be backed up by the sanctioning power of the parties involved in a particular transaction and/or the other actors in the community more generally. As a result of the above, they may have substantial anticipated effects on trust. If I am in a situation where appropriate informal institutions apply, and I am asked to trust someone who is subject to these institutions, I may reason that the informal institutions in question will apply across a variety of circumstances that I may only imperfectly be able to predict now. I may thus be more inclined to trust him on the basis that these informal institutions will provide a means of evaluating what trustworthy behavior is and ensure that he behaves trustworthily in these unanticipated situations.

Thus, we may expect that formal institutions will be associated with much "thinner" forms of trust than informal institutions. Precisely because formal institutions are better defined, less ambiguous, and more predictable, they will have relatively narrow and limited consequences for trust. This is not to say that social settings that are dominated by formal institutions will necessarily be low-trust. We may still expect to see trust and trustworthiness (or the lack of same) play an important role in the interstices of society, those areas of social interaction that are imperfectly, or not at all, covered by formal institutions. We may also see formal institutions being leveraged to make trust easier in those areas. Yet even if trust exists, it will not be socially important in the same way – actors will not need to rely on trusting relations in order to cooperate successfully across a wide variety of important social situations.

In contrast, informal institutions of the right kind[35] will be associated with much thicker forms of trust. They will provide practical guidance to actors about how to behave in situations that were not fully anticipated in the relevant institutional rules. If (and this is an important proviso) the informal rules in question are conducive to trust (viz., by stating that actors should cooperate in C-type situations), then we may expect that they will guide actors to behave in a trustworthy fashion, to believe that others will behave in a trustworthy fashion, and thus, in a very important sense, to trust each other in circumstances that were not specifically anticipated by the institution in question.[36] Again, this is not to say that settings dominated by informal institutions, rather than formal institutions, will necessarily be high-trust settings. But they will be settings in which trust-based relationships play an important role, as actors will, in the absence of institution-produced confidence, rely on relationships with those whom they can trust if they wish to have a reasonable expectation that their cooperation will be reciprocated. Thus, for example, even though they are both dominated by informal institutions, the high-trust environment among diamond merchants described by Lisa Bernstein (1992) is very different from the low-trust environment of Edward Banfield's (1958) Mezzogiorno village. Nonetheless, even in the latter setting, trust plays a key role – actors are limited to interactions with those whom they can trust,

[35] I discuss further some of the attributes of the "right kind" of institutions in Chapter 6.

[36] For more discussion of which informal institutions are appropriate, and which are not, see Farrell and Knight (unpublished).

primarily other family members.[37] These differences between formal and informal institutions will have important empirical consequences, a point that I develop at greater length in Chapter 5.

2.7 CONCLUSIONS

I have laid out an account of the relationship between institutions and trust. This account seeks to bridge – and synthesize – two apparently irreconcilable literatures and to mitigate three problems that plague current discussions of this relationship.

Most importantly, the arguments reconcile two accounts of cooperation, which both invoke the assumption that actors behave rationally, but which appeal to very different notions of how structures impinge on rational actors' interests. In the encapsulated interest account of trust, the important structures are personal relations, which provide both sufficient incentive for actors to behave trustworthily in many instances and sufficient knowledge to actors of others' incentives to allow trustworthiness (or, more precisely, the perception of trustworthiness) to be translated into trust. This account allows us to understand the circumstances under which a rich version of trust can flourish, even among rational actors, but it has great difficulty in incorporating the effects of institutions. However, dominant rational choice accounts of institutions give us a clear understanding of how institutions can have determinative consequences for cooperation – but not of how they can affect the rich kinds of trust that Hardin and others discuss.

By treating institutions as mechanisms of equilibrium selection, and by examining the ways in which they can guide actors to converge on certain equilibria by providing information on how others are likely to behave in ambiguous circumstances, it is possible to bridge these two literatures and, indeed, to synthesize them. This synthetic approach also helps mitigate three important problems in the existing literature on institutions and trust.

[37] See Cavalli and Cavazza 2001, 130: "where familism remains strong in Italy, it should be understood as an exaggerated distrust, more widespread in Italy than elsewhere, of the impersonal institutions of both the market and the state."

First, it avoids a functionalist account of institutional origins and change, in which institutional evolution is driven by functional imperatives such as, for example, the broad social need to underpin trust and cooperation. Such accounts are rife in the literature, but they very frequently assume what they should set out to prove (viz., that the institution was indeed created as a result of this functional need). Instead, it argues that institutions are only indirectly linked to cooperation; that is, that they arise from power struggles over how the benefits of cooperation should be distributed.

Second, it distinguishes clearly between cause and effect. It argues that institutions may be seen as an external selection mechanism, which guides actors to converge on certain equilibria rather than others. Appropriate institutions may thus guide actors to converge on equilibria in which they behave in a trustworthy manner (inappropriate institutions, obviously, may have quite the opposite effect). Even more importantly, institutions may give actors information on how others are likely to behave. Because institutions are effectively rules about how different types of actors should interact, they contain information about the different types of actors to be found in a society, and how these types may be expected to behave. This information may be valuable not only in situations that are directly and clearly covered by institutions but also in other situations that are only imperfectly or ambiguously covered.

Third, this account provides at least a partial remedy for the overly determinate accounts of the relationship between institutions and trust that plague the current literature. Trust necessarily involves some degree of uncertainty – in particular, it is best invoked in contexts where actors are not able to anticipate exactly what situations might arise in the future. To say that I trust you is to say that I believe that you will behave in a trustworthy manner in future circumstances that are not possible to anticipate (and thus to create, say, appropriate contractual mechanisms) at this point in time. An account of institutions, which stresses how they may provide guidance to actors in just such ambiguous future situations, provides a grounded account of how it is that institutions may affect trust (by providing me some reassurance that the trusted party will behave appropriately, even in situations that we cannot fully predict) without having that determinative effect on expectations that authors like Gambetta and Lorenz argue is incompatible with trust.

Finally, the more complex account of institutions and trust presented in this chapter not only allows us to understand how institutions may affect trust but also to generate hypotheses about the circumstances under which institutional variation and institutional change is likely to be associated with variation and change in the ways in which actors trust each other. First, it argues that institutions' effects on trust may change over time, as the institutions themselves change in response to shifts in power relationships. The range of issues over which one actor can trust another may narrow or broaden as power relations change. Second, it argues that different kinds of institutions will have different consequences for trust. In particular, we may expect informal institutions to have a much more systematic impact on trust, precisely because of their diffuseness – they will allow actors to trust each other over a broader set of issues that is more difficult to define beforehand. In contrast, formal institutions will have a much narrower impact on trust, although they may very well underpin widescale public confidence, exactly because their consequences are relatively predictable.

3

Varieties of Capitalism and Industrial Districts

3.1 INTRODUCTION

The previous chapter lays out a theory of the relationship between institutions and trust, showing how institutional variation is likely to be associated with variation in patterns of trust (and thus cooperation) and how institutional change over time will have knock-on consequences for trust. How is this relevant for the kinds of trust and cooperation that political scientists and other social scientists are interested in? There are many possible applications. The larger part of the remainder of this book takes up one of the more important ones, examining trust and cooperation in the context of comparative political economy and, more particularly, the literature on cooperation among firms.

Comparative political economy needs a proper theory of the relationship between institutions, trust, and cooperation. Over the last several years, scholars in this tradition have made some real theoretical advances. The "varieties of capitalism" literature (Shonfield 1969; Albert 1993; Hollingsworth, Schmitter, and Streeck 1994; Hollingsworth and Boyer 1997; Crouch and Streeck 1997; Hall and Soskice 2001) offers an account of how institutions matter to the economy and of how differences in institutions are likely to be associated with differences in both economic policy making and firm behavior. A recent and highly influential synthesis (Hall and Soskice 2001) seeks to apply rational choice arguments about institutions to the traditional concern of comparative political economy – explaining variation in the forms of coordination to be found in different economies.

This synthesis, while a major advance, is unlikely in its current form to provide a rigorous and complete account of the relationship between institutions and cooperation in the political economy. There are some important weaknesses in Hall and Soskice's microfoundations. As already noted in Chapter 1, Varieties of Capitalism theory grounds its explanation of political economy in strategic interactions among firms, but it does not provide a satisfactory account of the sources of institutions nor of institutional change. Instead, institutions are more or less exogenous structures that provide particular solutions to coordination problems. Thus, this account shortchanges another set of strategic problems – those involved in institutional creation (Riker 1980). Although Peter Hall and Kathleen Thelen have tried to supplement the varieties of capitalism approach with a theory of institutional change (Hall 2005; Hall and Thelen 2007), this account pays little attention to the kinds of informal institutions that are likely to be crucial to explaining trust (see further Chapter 5). Equally, the Varieties of Capitalism approach has only a vague theory of institutions' effects – it argues that institutions induce actors to coordinate for "normative, cognitive or material reasons" (Hall and Soskice 2001, 9). While this accommodates a very wide variety of possible institutional effects, it equally makes for a somewhat murky account of the relevant causal relationships, with a correspondingly limited capacity to predict variation in outcomes.

The theory of institutions and trust outlined in the previous chapter certainly does not provide a replacement for more general theories of the political economy, such as Hall and Soskice's. Many of the issues that concern comparative political economists do not involve trust and cooperation so much as coordination and the provision of collective goods through government or other means. For example, bilateral trust and cooperation among economic actors to create a collective good may be irrelevant in contexts where the actors can rely on an outside actor (such as the government) to produce the good for them (Farrell and Holten 2004). Thus, a theory of trust and cooperation alone is not enough to explain many aspects of the political economy.

Such a theory does, however, provide a more grounded and rigorous account of outcomes in those areas of economic activity where trust and cooperation among actors are relevant and important. Cooperation among firms is one highly important example. Much of the earlier literature on varieties of capitalism sets out to explain cooperation among firms, as does Hall and Soskice's account. However, the earlier literature lacks comparative bite (Zeitlin 1992), while Hall and Soskice's theory, for the reasons outlined above, has deficiencies in its accounts both of where institutions come from and what their precise consequences for cooperation might be. For this reason, neither is really well suited to provide a proper account of the "political economy" of trust. An account of the kind provided in the last chapter – which seeks both to provide a grounded account of institutional change and an account of institutions' particular effects on trust – offers a better alternative (and a complement to those other mid-range theories which explore institutional effects in settings where trust is not of primary importance).

The remainder of this book seeks to assess the merit of this theory, both by seeing how well it measures up in explaining a continuing anomaly in our understanding of cooperation among firms and by testing whether it can be extended to a setting where distrust rather than trust obtains. Patterns of trust and cooperation in so-called "industrial districts," geographically concentrated areas of small (and medium-sized) firm production, pose an important challenge not only to varieties of capitalism explanations but also to rational choice explanations of cooperation more generally. The overwhelming consensus in the existing literature is that the kind of trust and cooperation observed in industrial districts cannot be explained by frameworks that appeal to the rational actor model. In many industrial districts, firms seem to cooperate without formal contracts or backup from a well-functioning legal system, dividing up the production process among hundreds of small firms. This is not what conventional economic scholarship (Williamson 1975; 1985) would predict. Nor does the varieties of capitalism approach, as it is currently constituted, do much better; its focus on national-level institutions means that it has difficulty in explaining more localized phenomena such as industrial districts.[1]

[1] This said, it is necessary to pay attention to the interaction between national and local institutions. See Chapter 4.

Earlier, more sociological variants of the varieties of capitalism approach were heavily influenced by work on industrial districts. Indeed, academic work on the organization of production, which drew heavily on evidence from industrial districts, made a "fundamental contribution" to comparative studies of capitalism (Hall 1999, 9), leading scholars both to explore the role of regional and local organization[2] and to take an explicit interest in inter-firm cooperation (Hall 1999). It is not immediately clear whether the currently dominant version of the varieties of capitalism approach, which rests on a fundamentally rationalistic account of institutions, can do as well. Scholars of industrial districts disagree about the exact sources of cooperation, but they typically agree that nonrational phenomena are key to explaining it. Thus, the literature on industrial districts poses an explicit challenge to rational choice theories of trust and cooperation in an area – cooperation among firms in the political economy – where they claim to have considerable purchase.

The theory that I have outlined in the previous chapter provides one way in which scholars of the varieties of capitalism, and of rational choice institutionalism more generally, might respond to this challenge. This chapter and the three chapters following will assess whether this theory provides us with an understanding of cooperation among firms within these districts and indeed of the lack of cooperation other actors (members of the Sicilian Mafia).

The remainder of this chapter lays the initial groundwork for assessing this theory's empirical merits. It begins by outlining more precisely what industrial districts are and why they matter for the study of trust, trustworthiness, and cooperation, providing a very short overview of the existing literature and how it has evolved, with particular emphasis on the relationship between trust and cooperation in industrial districts. Next, it discusses two key cases within the industrial district literature – the packaging machinery industrial district in Bologna, the capital of the Italian province of Emilia-Romagna, and the machine tool industrial district in and around Stuttgart, the capital of Baden-Württemberg in Germany. Emilia-Romagna and Baden-Württemberg are two of the paradigmatic cases in the existing

[2] See, for example, Hollingsworth et al. 1994; Crouch et al. 2001; 2004; Crouch and Farrell 2004.

literature – they thus provide reasonable proxies for a more general phenomenon. To the extent that an institutionalist explanation can provide a good explanation of how trust and cooperation work within them, we may reasonably surmise that a sophisticated institutionalist framework is likely to provide insights into industrial districts more generally. This chapter provides short accounts of mechanical engineering in these two regions, providing necessary background information for the more detailed empirical discussion in the succeeding chapters. It then concludes by discussing whether cooperation in these districts is driven by interests or by other factors.

3.2 THE EXISTING LITERATURE ON INDUSTRIAL DISTRICTS

The existing literature on industrial districts is plagued by conceptual stretching (Sartori 1984). The term "industrial district" has been used to refer to types of geographic space, to sociological or cultural phenomena, and even to particular kinds of policies (DiGiovanna 1996). This plethora of definitions reflects both disciplinary differences (the concept has been taken up by sociologists, by geographers, by scholars of public policy, and by political scientists) and basic theoretical confusion. Scholars across a wide variety of fields of inquiry agree that industrial districts are important but disagree not only about *why* they are important, but about what they are.

I use a relatively narrow definition of industrial district. The term "industrial district," as I use it, refers to a geographically concentrated cluster of small manufacturing firms engaging in the same sector of production, or in closely related sectors, and benefiting extensively from economies of proximity. I examine the origins of one, albeit crucial, subset of these economies – the economies that are produced by cooperation between firms located within the industrial district in question.

Recent interest in these districts has its origins in the study of Italy's political economy and dissatisfaction with traditional, dualistic approaches (e.g., Gramsci 1971) that stressed the difference between the underdeveloped, agrarian South and the industrial North. In this account, firms in the South were seen as small, inefficient, and locked into local markets, whereas northern firms were larger, more economically successful, and internationally oriented. As a group of writers began to point out in the 1970s, this dualistic approach failed to explain a third pattern of economic development, to be found in the center and northeast.[3] In the so-called "Third Italy" (Bagnasco 1977) localized systems of small-firm production appeared entirely capable of achieving success on export markets and indeed of capturing many of the efficiencies available to large firms.

Italian scholars sought to understand this phenomenon by turning to Alfred Marshall's work on how "industrial districts" of small producers or "masters" in nineteenth-century Britain seemed able to capture many of the economies of scale more usually associated with large firms.[4] Marshall had attributed these districts' success to the rather vaguely defined phenomenon of "industrial atmosphere." Italian political economists and sociologists tried to find more specific explanations for the sources and likely prospects of these districts. In many cases, although certainly not in all, the districts seem to have had their beginnings in the decision of larger firms to subcontract out work. Following the *autunno caldo* (hot autumn) of social unrest in 1969, large companies were forced to make major concessions to their workers, including the establishment of work councils and the so-called *scala mobile* system of wage indexation.[5] Many of them responded by outsourcing stages of the previously in-house production process to smaller firms, which found it easier to avoid unionization and carried a lighter regulatory burden. This led many scholars to argue that industrial districts were a contingent by-product of large firm strategies. Others countered by arguing that industrial districts had capitalized on changing market conditions that favored smaller, specialized producers (Brusco 1982) and by showing that the patterns of cooperation observed was not a simple product of the outsourcing strategies of large firms (Trigilia 1986). These controversies were accompanied by a series of case studies of individual industrial

[3] For critical appraisals of dualist approaches to Italian society and economic development, see Bagnasco 1977; Bagnasco and Messori 1978; and Brusco 1990.

[4] For a brief overview of Marshall's thought, which emphasizes its contribution to the study of the economics of geography and space, see Krugman 1991.

[5] For a useful brief summary of the development of Italy's industrial relations system, see Chapter 3 of Locke 1995; also see Regini 1983; 1995; and Regini and Regalia 1997.

districts that supplemented macrolevel arguments about the changing nature of industrial production with a specific focus on the patterns of cooperation within industrial districts that allowed them to replicate many of the advantages more usually associated with large firm hierarchy.[6]

This debate attracted international attention after the publication of Michael Piore and Charles Sabel's (1984) influential *The Second Industrial Divide*. Piore and Sabel depicted industrial districts as the harbingers of "flexible specialization," a trajectory of production involving small, highly flexible firms supported by local institutions, which provided a viable alternative to the mass production techniques of traditional Fordism. Although *The Second Industrial Divide*'s arguments were highly controversial, they led researchers to look for industrial districts outside Italy.[7] Studies of small-firm clusters in Germany, Denmark, Spain, the United States, and elsewhere began to use the concept of the industrial district, most prominently including Gary Herrigel's series of studies of the machine industry in Baden-Württemberg.[8] However, as the concept came to be applied by scholars outside its original context, it also began to stretch, exposing basic limits in the conceptual model (Zeitlin 1992). Furthermore, the problems that many industrial districts encountered in the early 1990s gave impetus to a revisionist literature, which depicted them as either largely epiphenomenal to the global economy (Amin 1993, Amin and Thrift 1992; 1994) or destined to be steamrollered by large firms (Harrison 1994). This body of work has in turn set the stage for a new set of studies examining how industrial districts have responded, with more or less success, to changing economic conditions (Bagnasco 1995, Herrigel 1999, Burroni and Trigilia 2001, Whitford 2001a, Whitford and Potter 2007, Berger and Locke 2001). Although it is clear that industrial districts have not provided a generalizable

[6] See Trigilia (1989) for an exemplary treatment of an important industrial district, the textile producing district in Prato, near Florence.

[7] The literature responding to *The Second Industrial Divide* is extensive, complicated, and for the most part only marginally relevant to the issues under discussion here. Important contributions include Piore (1992), Hirst and Zeitlin (1992), Jessop (1992), Amin (1993; 1994), Amin and Thrift (1992; 1994), Scott and Storper (1992), and Storper (1995).

[8] See especially Herrigel 1996a; also see Herrigel 1990; 1993; 1996b; 1999.

model of economic development, many of the "classic" industrial districts seem to have weathered changing economic conditions with a reasonable degree of success, suggesting that they are not necessarily transitory or epiphenomenal as some of their harsher critics have suggested.

The way in which the literature has developed – from the study of a specific set of cases in Italy to a more general body of cross-national research – has had two unfortunate consequences. First, scholars have had difficulty in creating appropriate abstractions that would allow them to properly compare different kinds of industrial district in different contexts. In Jonathan Zeitlin's (1992) description, the literature on industrial districts has had difficulty in moving from a "thick" model based on a few *loci classici* in Italy, to a "thin" one that might provide a better explanation of variation across cases. The result has been a dearth of careful cross-national comparative studies in the literature – most work on industrial districts has consisted of single case studies or, at most, of quasi-comparative studies within a particular regional or national context.[9]

Second, much of the literature is imbued with an implicit or explicit set of value judgements – that Italian industrial districts present a normatively superior alternative to conventional forms of mass production. While this bias is defensible, it has sometimes tinged the literature with a small-firm romanticism that has clouded judgements about empirical reality (Bagnasco 1995), and thus has made it more difficult to arrive at sound conclusions both about the factors affecting industrial districts and their broader social consequences.

3.3 INDUSTRIAL DISTRICTS, COOPERATION, AND TRUST

I do not aim to provide a general theory of industrial districts as an economic phenomenon. My intent is much more specific – to provide a theory of the kinds of trust underpinning diffuse cooperation among firms within these districts. It now appears unlikely in the extreme that industrial districts are going to provide a new, generalizable model of

[9] Important exceptions include Cooke and Morgan (1998), which is mostly concerned with differences in policy, Crouch et al. (2001; 2004), and Courault and Romani (1992).

production, but they still present an important challenge to our understanding of cooperation in the political economy. Sebastiano Brusco (1982) sums up the findings of a broader literature when he claims that they involve a mixture of "competition and cooperation," where firms compete vigorously with each other on some aspects of production, but are willing to collaborate with each other where necessary to produce the final product. What explains this mixture? When are firms likely to cooperate, and when are they likely to compete with each other? More generally, what are the sources of trust and cooperation in industrial districts?

To answer these questions, it is useful first to distinguish between two kinds of cooperation within industrial districts: vertical cooperation and horizontal cooperation. Vertical cooperation – that is, cooperation in the production process between final firms (which sell the finished product on the final market, and perhaps design it), their subcontractors, and sub-subcontractors – is arguably the more important of the two. Many industrial districts have an extraordinary level of disintegration of the production process, so that tasks usually undertaken within the firm hierarchy are carried out through vertical cooperation between small independent producers. They thus appear to solve problems of opportunism without resort to vertical integration.[10] Production in these districts is carried out through cooperation among a myriad of small firms. Although some firms play an important role in organizing the production process and in providing an interface between the district and the final markets for the goods in question, they do not act as hierarchical superiors in the normally understood sense of the word.

Horizontal cooperation – cooperation between firms that do not directly buy from, or sell to, each other, but instead offer complementary or competing production inputs within industrial districts – is a second important characteristic of many industrial districts. Firms may cooperate horizontally, for example, through forming consortia to buy or sell products in common. They may exchange information on possible customers or technical problems. The literature suggests that horizontal cooperation among firms is very common in industrial districts (Pyke and Sengenberger 1992; Herrigel 1996a), although it notes

[10] I discuss this further in Chapter 4.

that direct competitors are usually disinclined to cooperate with each other (Brusco 1982). This aspect of cooperation has been highlighted in much of the public policy literature on industrial districts; both government and firm associations have historically sought to support horizontal cooperation among firms through targeted policy initiatives (Brusco 1982; but see also Farrell and Holten 2004). However, some recent work (Innocenti 1998; Cooke and Morgan 1998) suggests that the importance of horizontal cooperation has been significantly exaggerated. In later chapters, I discuss vertical cooperation between final firms and subcontractors at greater length, since such cooperation is more easily comparable across contexts than other varieties.

The prevalence of cooperation in industrial districts poses three important challenges to rational choice accounts of trust and cooperation in the economic sphere. First, and most obviously, where does this cooperation come from? Some scholars of Italian industrial districts suggests that it is the product of surviving cultural traits from Italian peasant society (Becattini 1994) or of historically rooted patterns of reciprocity (Putnam 1993). Others point to the importance of local networks of political associations (Trigilia 1986) and of patterns of political mobilization dating from the late nineteenth century (Bagnasco and Trigilia 1993). Still others attribute trust and cooperation in industrial districts to local rules (Bellandi 1992; Brusco 1992) or to the dynamic reformulation of collective histories and identities (Zeitlin 1992; Sabel 1993). With very few exceptions (Lorenz 1992b; Dei Ottati 1994; Farrell 2005) they privilege nonrationalistic factors such as culture, identity, and discourse as the key sources of trust and cooperation.

There is some force to these claims – rational choice only provides a partial and limited account of human motivations. While some kinds of cultural framing can be incorporated within rationalistic accounts, they are difficult or perhaps even impossible to model using game theory (Johnson 2002). Even so, dominant approaches to the explanation of industrial districts arguably overdiscount the role of considered self-interest in explaining the degree and kind of economic cooperation that occurs within them. As a result, they may thus be missing an important part of the explanation for the degree and kinds of vertical and horizontal cooperation that can be observed in these districts.

Thus, the first challenge to rationalistic accounts of cooperation in industrial districts is quite straightforward: to establish whether self-interest plays an important role in explaining trust and cooperation. The encapsulated interest account of trust described in the previous chapter provides one theory of how trust may flow from self-interest. Under this account, I trust you to the extent that I believe you have an interest in behaving in a trustworthy fashion. I seek to assess what your interests are, and then, on the basis of that assessment, I may trust you or distrust you. If this account has substantial explanatory value, we may expect that actors' decisions to trust others will depend on their beliefs regarding those others' interests. We may also expect that there will be a strong relationship between interests and trust. Thus, we may expect that where actors have clashing interests – that is, the interests of the trustee seem at odds with the interests of the truster – trust will be difficult to achieve. In contrast, where the interests of the trustee and the interests of the truster seem consonant with each other, trust will be relatively easy to establish. In principle then, the encapsulated interest account should provide a differentiated explanation of cooperation. Where actors' interests are consonant with each other, they will have a strong incentive to cooperate over matters of mutual convenience. Where actors' interests, in contrast, clash with each other, they will not cooperate.

The second and the third challenges that industrial districts pose to political economy flow from the need to explain variation over both space and time. The existing literature on industrial districts does a remarkably poor job in explaining both kinds of variation. It consists for the most part of single case studies, with little effort to use controlled comparison in order to explain variation in space. Discussions of change over time in industrial districts have tended to have weak microfoundations, relying on secular shifts in forms of social organization as their main explanatory variable (Franchi and Rieser 1991; Harrison 1994).

The encapsulated interest account arguably provides exactly the kind of microfoundations that are necessary for a more finely grained account of variation in space and time. However, in order to provide such an account, it is necessary to provide a somewhat narrower account of trust and cooperation. In addition to arguing that trust and cooperation are driven by interests, it is necessary to specify more clearly *which* kinds of interest are likely to matter. As discussed above, the encapsulated interest account of trust has some predictive power. However, it is possible that actors' interests may flow from a variety of sources, including institutional incentives, networks, and normative commitments. In order to provide tighter predictions regarding variation in kinds of trust and cooperation and change over time, there is good reason to narrow the focus to institutional factors. Clearly, this will miss out on some of the factors that define actors' interests, but in the specific case of inter-firm cooperation (which is supported by trust among business actors) there is good reason to believe that institutions will play a particularly important role. While relationships of friendship may help cement business relationships, and kinship ties have proved a key guarantor for business transactions in earlier eras of history (Greif 1994), most economic relationships in advanced industrialized democracies will involve shallower ties between individuals who do not necessarily have strong personal relationships with each other. In such circumstances, institutions – whether formal or informal – are likely to provide a compelling source of information about the likely interests of others. The theory outlined in the previous chapter seeks to provide exactly an account of how institutions will affect trust and cooperation through providing such information.

3.4 TWO INDUSTRIAL DISTRICTS: MACHINE PRODUCING IN BOLOGNA AND STUTTGART

Two of the classic cases in the industrial district literature, the packaging machinery district in Bologna, the capital of the Italian region of Emilia-Romagna, and the machine tool district in Stuttgart, capital city of Baden-Württemberg, provide an excellent testing ground for theories of the relationship between institutions, trust, and cooperation. First, they are among the most important cases in the existing literature. The industrial districts of Emilia-Romagna are arguably, together with the textile-producing district of Prato in Tuscany, the most widely cited and studied examples of the phenomenon in Italy.[11]

[11] For a sampling of this literature, see Bellini et al. 1990; Bianchi and Gualteri 1990; Brusco 1982; Brusco et al. 1996; Cooke and Morgan 1994; 1998; Heidenreich 1996; Leonardi 1990; Nardin 1993; Perulli 1989.

Equally, machine production in Stuttgart and in other parts of Baden-Württemberg is perhaps the most widely studied example of an industrial district outside Italy.[12] The existing literature claims that trust and cooperation in industrial districts are nonrational phenomena. Thus, these should be "hard cases" for the rational choice approach. If the encapsulated interest account provides a good explanation of trust and cooperation in these paradigmatic cases, it suggests that this account likely has considerable explanatory power elsewhere.

Even more to the point, the two districts provide good cases for comparison using a "most similar systems" or "comparable cases" design (Przeworski and Teune 1970, Lijphart 1975). They minimize variation on important extraneous variables. They involve production in closely related sectors – the production of packaging machines and machine tools. This is not to say that there are not some significant differences between the two cases. Machine tool producers tend to be more technologically advanced than packaging machinery producers and to rely more on expensive research and development. Many machine tool producers in Stuttgart sell to a local customer base in the automotive sector, while packaging machinery producers in Bologna rely heavily on export markets. Even though these differences mean that it is difficult to make controlled comparisons of how institutions might affect cooperation in research and development, or cooperation between producing firms and their customers, manufacturers in both districts face highly similar problems of how to organize cooperation and production. Both need to manufacture complicated machines that require a large number of subcomponents. Both, accordingly, face a choice between putting out the manufacture of these subcomponents to specialized subcontractors or manufacturing them in-house. As discussed in Chapter 4, this provides a reasonable degree of control, allowing us to analyze the different ways in which machine manufacturers might choose to cooperate, or not cooperate, with other firms.

[12] See for example Amin 1994; Braczyk 1996; Braczyk and Schienstock 1996; Braczyk, Schienstock and Steffensen 1996; Cooke and Morgan 1994; 1998; Grotz and Braun 1997; Heidenreich 1996; Herrigel 1993; 1996a; 1996b; 1999; Heidenreich and Krauss 1999; Semlinger 1993a; 1995; Shmitz 1992; Staber 1996.

Finally, the cases have a substantial degree of variation on the key independent variable of interest – institutions. This claim is developed further in Chapter 4; for now, suffice it to say that Germany's well-developed system of national, formal institutions provides an important contrast with Italy's political and economic system, in which local, informal institutions tend to predominate. If the arguments developed in Chapter 2 are correct, we may expect that these differences will be reflected in important differences in patterns of trust and cooperation between the two cases.

In order to investigate patterns of trust and cooperation in the two cases, a body of detailed, semistructured interviews were conducted in both Bologna and Stuttgart over 1998 and 1999, a period when machine manufacturing in both localities was undergoing important change. Interviews in Italy were conducted either by the author or by Ann-Louise Lauridsen with the author present. Interviews in Germany were conducted by Ulrich Glassmann with the author present in most cases. These interviews were conducted with firms in the relevant industries and with local actors (business organizations, academic research institutes, local/regional/state government officials) who might be expected to provide more general insights into the functioning of the two local economies. Seventeen firms were interviewed in Bologna, as well as 23 nonfirm actors with knowledge of the local economy. Twenty firms were interviewed in Stuttgart, as well as nine nonfirm actors. To preserve confidentiality, the Italian firm interviews have been numbered (1–17), and the German interviews (18–37). Translations from Italian and German were carried out by the author.

3.5 EMILIA-ROMAGNA AND THE PACKAGING MACHINERY INDUSTRY IN BOLOGNA

The Italian region of Emilia-Romagna plays a key role in the existing literature on industrial districts.[13] Districts in this region include the machine manufacturing district in Bologna, the agricultural machinery cluster in Reggio-Emilia, the ceramics (and ceramic machine) producing district of Modena, the food and food processing machinery

[13] See Amin (1998) and Farrell and Holten (2004) for more detailed discussions of changes in the "Emilian model."

cluster in Parma, the medical instruments cluster at Mirandola, and the garment producing cluster centered on Carpi.

Bologna is the capital city of the region – its surrounding metropolitan area, together with that of the smaller city of Imola, make up the province of Bologna.[14] The province of Bologna has roughly 900,000 inhabitants – like many areas of northern Italy, it has suffered from demographic decline over recent years. Bologna has a strong manufacturing sector, which accounts for just under 30% of total private sector employment, with machine building and metalworking accounting for just over half of all manufacturing employment (ISTAT 2001 industrial census). Thus, while the machine industry does not dominate Bologna as, say, textiles do the industrial district of Prato, it plays a very substantial role in the city's economy. This role goes beyond the employment that it provides – the specialized packaging machinery industry was the main motor force driving Bologna's economic success in the postwar period (Capecchi 1997). Within the machine industry, packaging machinery, and related sectors (tobacco processing machines) play a key role. The Italian packaging machinery industry had turnover of 2.9 billion Euro in 2003; over 80% of Italian packaging machinery firms are located in Emilia-Romagna, in a densely populated cluster centered on Bologna and the nearby city of Modena.

In contrast to some other Italian industrial districts, the packaging machinery cluster in Bologna has relatively recent origins. Before World War II, there were only two local companies in the sector.[15] After the war, production expanded rapidly as new markets began to open up. Not only did these two firms, and a third that had entered after the war, grow rapidly in size, but they gave birth to a host of newer firms, as technicians who they employed left to start up their own businesses. Some of these firms in turn gave birth to new ones.

[14] In the parlance of Italian administration, the province is an intermediate level of geographic aggregation between the region and the commune. While the provincial level of administration is of little interest in itself, ISTAT (Istituto Nazionale di Statistica), the Italian national statistics agency, provides extensive data that allow for the examination of change in industrial structure over time (in contrast to other levels of aggregation, such as *sistemi locali* or travel-to-work areas, for which data is only patchily available). For reasons of consistency, I use provincial-level data throughout.

[15] See Chapter 5.

As the mechanical engineering industry grew in sophistication, firms were able to enter – and in some cases create – new market segments. Patrizio Bianchi has argued that informal local rules exist in the Bologna packaging industry, as in other such local economic systems, such that firms agreed to make complementary rather than competing products.[16] The historical evidence seems at best to lend heavily qualified support to this assertion: while "daughter" firms usually did not compete with their mothers, and sometimes received help from them, it was by no means unknown for daughter firms to make products that competed directly with those of the firm which had given them birth (Capecchi 1997).

The industry saw continued expansion during the 1950s and 1960s as export markets continued to grow – the worldwide boom in manufacturing led to ever-expanding demand for packaging machines. The 1950s also saw many employees leaving their firms under less than friendly circumstances, as difficult labor relations led to mass firings. Relations between firms and employees became ever more difficult during the 1960s, culminating in the *autunno caldo* of industrial unrest.

Growth continued during the 1980s: overall employment in specialized mechanical engineering in Bologna jumped from 8,296 in 1971 to 12,451 in 1981, afterwards declining to 11,354 by 1996, along with manufacturing employment more generally in the region.[17] Growth in turnover continued during the 1990s, reaching a peak of 20.3% between 1994 and 1995, but declining in the late 1990s because of turbulence in final markets and a drop in demand from traditional customers (large cigarette producers). By 1999–2000, the Bologna packaging machinery industry had reached maturity, with a relatively stable pattern of long-term growth, which it has continued to enjoy in the succeeding period (UCIMA [Unione Costruttori Italiani Macchine

[16] Bianchi 1993. See also Curti and Grandi 1997.

[17] Note that national employment in the packaging machine industry appears to have increased by 7.1% between 1994 and 1995, and 6.6% between 1995 and 1996, although it fell again by 2.5% between 1997 and 1998. See UCIMA 1998; 1999. Although packaging firms in Emilia-Romagna dominate the Italian packaging machine industry, it is hard to extrapolate downwards from national level aggregate data on employment, given the particular problems faced by Emilian firms in finding qualified workers during the 1999–2000 period when research interviews were conducted.

Automatiche per il Confezionamento e l'Imballaggio] 2003), although it continues to face challenges both in training personnel (Malusardi 1998) and incorporating new technology.[18]

Like many other Italian industrial districts, the packaging machinery cluster in Bologna is characterized by a radical disintegration of the production process, in which small firms and subcontractors play a key role. Figure 3.1 provides a breakdown of firms by numbers of employees in specialized machine manufacture in the province of Bologna.

As is immediately apparent, small firms play a highly important role: 458 firms in the district have less than 50 employees. Such firms account for 32% of total employment in the specialized machinery sector.[19] The Bolognese packaging machinery district has a complicated internal ecosystem in which a couple of very large firms, which combine subcontracting with an extensive capacity to produce components internally, coexist with a large number of small-to-medium size firms, which rely almost entirely on a vast array of subcontractors to build the final product. As Chapter 5 discusses, these subcontracting relationships involve extensive vertical cooperation. Historically, the district has seen a substantial degree of horizontal collaboration too, between firms building complementary machines that handle different parts of the packaging process.

Recently the packaging machinery district has seen important changes in patterns of ownership. Many specialized firms have been bought out and have become part of group firms, which seek to provide horizontally integrated packaging solutions for large customers. For example, a customer firm in the pharmaceutical industry might previously have bought a pill-dosing machine from one producer, a blister-pack machine from another, and cartoning and palletizing machines from other producers still. Now it is more likely to buy from

[18] Interview with Vittorio Capecchi, 1999.

[19] Source: ISTAT 2001 Industrial Census. This figure is almost certainly an underestimate, as it does not include many small firms, which carry out less specialized forms of subcontracting in the sector. Many artisanal firms which act as subcontractors to the larger firms are likely to be categorized by the statistics more generally as machinery producers or metalworkers, where they do not indeed seek to avoid official attention (and taxation) altogether.

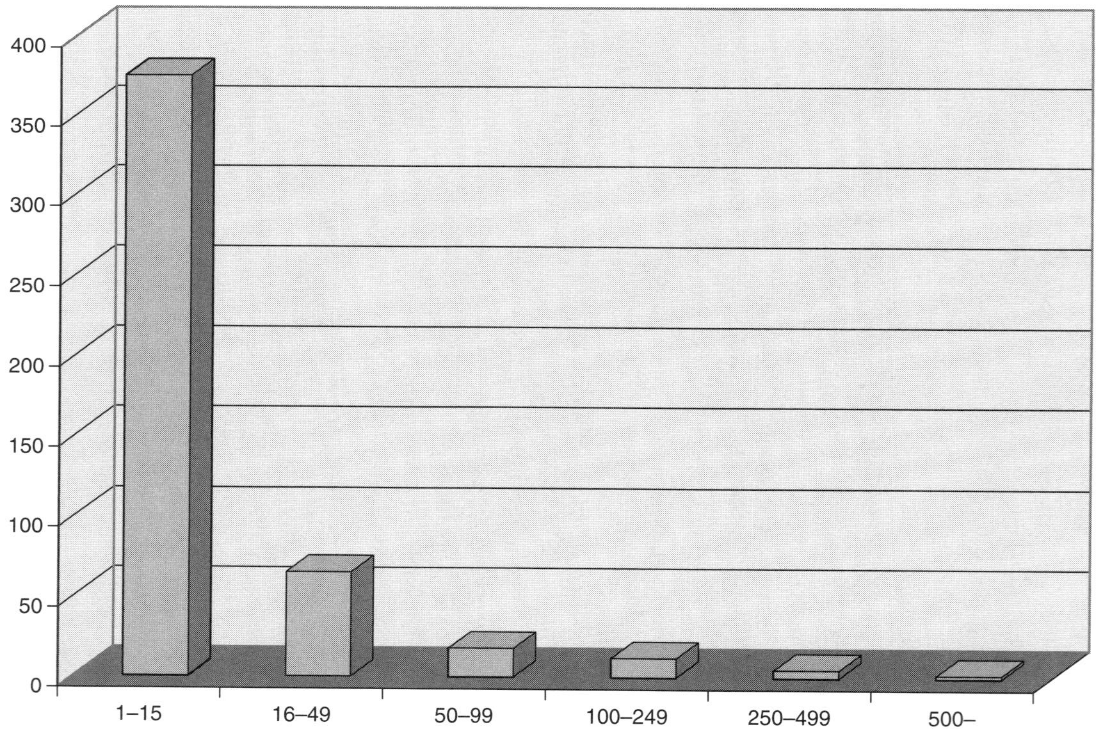

FIGURE 3.1 Number of firms in the province of Bologna classified by number of employees. Based on 2001 ISTAT industrial census data.

just one semi-integrated group firm, which has specialized producers for each type of machinery.

3.6 THE MACHINE-TOOL INDUSTRY IN STUTTGART, BADEN-WÜRTTEMBERG

The state of Baden-Württemberg has a population of 8.896 million, of which the region of Stuttgart accounts for over 25%, with a population of 2.27 million people. Like the province of Bologna, the region of Stuttgart contains the territorial capital and most of its surrounding conurbation.

Baden-Württemberg, like Emilia-Romagna, has received an extraordinary amount of attention from scholars working on industrial districts. This was in large part due to its economic success in the 1980s – many saw it as an example of how enlightened industrial policy and flexible manufacturing practices might combine to produce economic growth. Baden-Württemberg was often taken in the literature as a "model region" for economic development in the rest of Germany, while bargains between the Baden-Württemberg branch of the IG-Metall trade union and metal manufacturers in the region have traditionally set the pace for national level pay agreements.[20] However, the economic downturn the region experienced in 1992 and 1993 has seen a reassessment in the academic literature both of Baden-Württemberg's characteristic mode of internal organization and the degree to which regional policy initiatives (Heinze 1994) have been successful in promoting economic growth.[21]

[20] See Thelen and van Wijnbergen (2003) for a recent overview of trade union-business relations. For a specific discussion of the machine industry, see Müller-Jentsch et al. (1998).

[21] The early literature on Baden-Württemberg was largely positive in tone; see Piore and Sabel 1984; Sabel et al. 1989; Herrigel 1993; but also note the more cautious assessment in Shmitz (1992). Many recent assessments of Baden-Württemberg's regional path of development have been rather more guarded, if not pessimistic, and tend to argue that the market structures and patterns of interaction, which previously allowed for success have come under threat (Herrigel 1996a; Cooke and Morgan 1994), that Baden-Württemberg was never all that special in the first place (Grotz and Braun 1997; Staber 1996; Glassmann and Voelzkow 2001); or are undecided between these two positions (Braczyk 1996). For an important recent assessment, which argues that the academic literature (both positive and

In Baden-Württemberg as a whole, machine construction is the most important branch of industry, and it accounted for some 75 billion Deutschmark and 264,000 employees in the mid-1990s (Batz et al. 1998). Detailed statistical information on machine tool production in Stuttgart is not available; German data protection law prevents the publication of data that might inadvertently reveal firm-level information. However, the data indicates that there were 452 firms engaged in the more general sector of machine production in Stuttgart in 1999 (Glassmann 2004). Stuttgart is less dominated by its machine building sector in relative terms than many other regions in the state; indeed, it is unusual in that machine production is not the most important branch of industry in terms of employment (although it does employ 18.8% of workers in manufacturing; see Fuchs and Wassermann 2004). In absolute terms, however, Stuttgart's machine industry has a strong presence in terms of numbers employed and number of firms (see below); indeed, in the early 1990s, it accounted for some 33% of Baden-Württemberg's overall sales in mechanical engineering (Semlinger 1995). Baden-Württemberg also has a strong general presence in machine tool production – it accounted for an estimated 50% of total West German employment in the sector in 1990 (Müller-Jentsch et al. 1998). Machine tool firms in Stuttgart are almost certainly more dependent on the local vehicle manufacturing industry than their equivalents elsewhere in the state. This said, some commentators have argued that mechanical engineering firms in Baden-Württemberg, unlike those in Emilia-Romagna, are organized into production clusters on the level of the region rather than the locality.[22]

Machine building in Stuttgart has deeper historical roots than in Emilia-Romagna. While Baden-Württemberg industrialized rather later than many other parts of Germany, the first machinery firm in the (then independent) state was founded in 1852, and Baden-Württemberg had a well-established mechanical engineering

negative) has been unduly influenced by short term variations in economic success, see Glassmann (2004).

22 Heidenreich 1996. This is consonant with Gary Herrigel's (1996a) argument that the whole of Baden-Württemberg can be considered as an industrial district, although one should note that Herrigel's definition of an industrial district goes considerably beyond simple clustering effects.

industry by the end of the nineteenth century. Gary Herrigel (1996a) argues that small firms in regions of "decentered industrialisation" (such as Württemberg) were favored by the development of regional institutional infrastructures of support and by an interweaving of social and business relations among specialized producers, which fostered nonantagonistic relations among them (Herrigel 1996a). Firms specialized and engaged in complex patterns of subcontracting among each other in order to take advantage of each other's specialized knowledge: "collective experience, communication, and learning, led to mutual recognition, cooperation, and industrial self-regulation" (Herrigel 1996a, 175–76). In Herrigel's account, these arrangements came under pressure in the first three decades of this century, when changed market conditions led to cutthroat competition, which in turn undermined cooperation among producers. In the machine industry, firms sought stability through cartelistic arrangements, arranged through the trade association, which allowed member firms to specialize in particular niches, without having to fear competition from other firms. After World War II, some larger firms withdrew from this system of decentralized production, and cartels were banned (albeit with exceptions that allowed a wide variety of inter-firm agreements). Despite these changes, many small- and medium-sized firms were able to flourish in areas such as machine tool production where large firm industrial activity had created market niches for small, flexible suppliers. Many machine tool firms in Baden-Württemberg and Stuttgart succeeded in becoming world leaders in their particular niches; typically, they tended to dominate the high end of the market, providing superior quality, albeit at a price. Some machine tool producers, most notably Trumpf, succeeded in becoming relatively large firms, employing thousands of workers. Although the 1980s was a boom period for the machine tool and specialized machinery industries, machine tool producers in Baden-Württemberg experienced a severe economic crisis during the early 1990s.[23] Explanations for this crisis include inflexibility in the internal organization of the firm (Herrigel 1996b), the pressures of German reunification (Batz et al. 1998), unfavorable exchange rate and bank lending policies

[23] For detailed discussion, see Glassmann 2004.

(Glassmann and Voelzkow 2001), and various forms of lock-in (Cooke and Morgan 1998).[24] The Stuttgart region, which was especially dependent on the machine industry, saw particularly serious losses: between 1990 and 1997 the region lost more than a quarter of the jobs in the machine building sector (Batz et al. 1998). During the period that research was conducted, the industry had begun to enjoy a modest (albeit jobless) recovery, which has continued and indeed gathered pace, as firms have adapted to changed market circumstances and found new opportunities for technical collaboration and development (Glassmann 2004).

There is controversy among academics as to how the machine tool cluster centered around Stuttgart, and the machine industry of Baden-Württemberg more generally, is organized. On one side of the debate, Charles Sabel and Gary Herrigel have analyzed Baden-Württemberg in terms of the industrial district literature, stressing the degree of cooperation that exists among firms. On the other side, skeptics have argued that Baden-Württemberg is by no means unusual in terms of the degree of horizontal and/or vertical cooperation to be found among enterprises (Kerst and Steffensen 1995; Fuchs and Wassermann 2004).

Gary Herrigel, both together with Charles Sabel and separately, has characterized Baden-Württemberg as an industrial district. In his account, the economic success of the state was linked to the existence of dense horizontal and vertical networks of firms that cooperated in such a way as to allow each individual firm a considerable degree of

[24] Gary Herrigel (1996b) suggests that when faced with competition in high market segments from more flexible Japanese and U.S. competitors, German firms could not respond adequately because of rigid internal skill demarcations. Batz et al. suggest that this was paradoxically exacerbated by a boom in domestic demand resulting from reunification; just at the time when machinery producers were coming under increased pressure on international markets, they were lulled into a false sense of security by booming domestic markets. Glassmann and Voelzkow (2001) point to unfavorable rates of exchange and the refusal of banks to extend credit to many firms in difficulty as factors that helped exacerbate the crisis. Cooke and Morgan (1998) go on to identify three forms of lock-in which hampered Baden-Württemberg's governance system – functional lock-in, due to the high degree of reliance on vehicle manufacture, mechanical, and electrical engineering; cognitive lock-in, reflected by the slowness of firms to adapt to the changed market conditions that Herrigel identifies and political lock-in to the extent that government innovation policies are designed to support the traditional core sectors.

specialization. In this account, small- and medium-sized firms' ability to collaborate with subcontractors and peers was key to their success in the 1980s.[25]

Other scholars have disputed these findings. Cooke and Morgan found little evidence of institutionalised horizontal cooperation between firms.[26] Indeed, they report 60–70% of firms as arguing that there would be no need to collaborate if the government would only leave them alone.[27] Udo Staber (1996) reports on the basis of some two dozen interviews with business managers that horizontal collaboration does not figure prominently among firms in Baden-Württemberg. Others suggest that this dearth of horizontal cooperation is especially marked in the Baden-Württemberg machinery industry. Klaus Semlinger finds that

> in contrast to the close cooperation with their customers for which German machine-building firms are famous (vertical cooperation), horizontal cooperation in this industry has traditionally been only of minor significance and only recently – since the industry has faced severe problems – have corresponding ideas received some attention. (Semlinger 1995, 24)

Heidenreich and Krauss express a similar view when they argue that "cooperation between competitors . . . plays a relatively insignificant role in Baden-Württemberg"; they also find that the percentage of Baden-Württemberg mechanical engineering firms that report cooperative relations is only marginally above the German average (Heidenreich and Krauss 1998, 194). Grotz and Braun (1997, 547) find, on the basis of a survey covering mechanical engineering in three German regions (including the Neckar-Alb region of Baden-Württemberg) that "in all survey areas there is relatively little co-operation between competing firms." However, one should note that there has recently been some convergence between authors; Cooke and Morgan (1998) now accept

[25] Herrigel's later, more pessimistic assessment of machine producers in Baden-Württemberg sees their difficulties as primarily originating from internal inflexibilities rather than their external relationships (Herrigel 1996b). More recently, he has suggested that processes of "de-regionalisation" and "re-regionalisation," in which the nature of regional ties of cooperation between firms is changing fundamentally, is likely to have adverse and perhaps fatal consequences, for the previously existing industrial order; see Herrigel 1999.

[26] See further the summaries of the debate in Shmitz (1992) and Staber (1996).

[27] Quoted in Staber 1996.

that horizontal cooperation exists between some firms in Baden-Württemberg.

There is less controversy over the existence of vertical relations between machine producing firms and their customers. There is a high degree of consensus in the literature that deep collaborative relationships have traditionally existed between machine manufacturing firms and their customers. This is particularly so for machine tool producers in Stuttgart; very frequently their most important customers are the large car manufacturers of the locality. The industrial structure, in short, is one where the final buyer firms (often large car manufacturers or their system suppliers) play a central role (Shmitz 1992; DiGiovanna 1996; Heidenreich 1996. This contrasts with the packaging machinery industry in Emilia-Romagna, where, even though there is an important recent trend toward closer long-term relationships between final customers and machinery firms, such relationships still are looser than in Baden-Württemberg. These differences are tied into broader differences in the organization of production.[28]

There is also disagreement as to the extent of cooperation in vertical relationships between machine-producing firms and their subcontractors, as opposed to their customers. While both Herrigel, on the one side, and Cooke and Morgan, on the other, believe that such cooperation is important, several other academics are more skeptical. Semlinger (1995), for example, finds that buyer-supplier relations often resemble exploitative "henchmanship" rather than genuine cooperation. Staber (1996) suggests that there is little solid evidence that inter-firm cooperation of any sort is important in the region, and Grotz and Braun (1997, 549) find that "local sub-contractors mainly perform low-level production operations" of a sort that offers little scope for complex collaboration. Further, some observers note that machine tool firms in the region are typically highly vertically integrated and

[28] In Cooke and Morgan's argument, "to the extent that there are distinctive inter-firm relationships of long standing [in Baden-Württemberg], they are structured in *vertical* supply chains which often link the *Mittelstand* to large firms at the centre of elaborate inter-firm networks. This, it can be argued with reasonable confidence, is what differentiates the industrial fabric of Baden-Württemberg from that of Emilia-Romagna where lateral and vertical relationships among small firms in tightly defined industrial districts are quite commonly found" (Cooke and Morgan (1998, 101).

reluctant to subcontract out important parts of the manufacturing process (Semlinger 1993a; Heidenreich 1996). In contrast to other firms in the metal industry (Thelen and van Wijnbergen 2003), Baden-Württemberg machine producers have not typically extended their supply chains for important components to cheaper manufacturers in the former GDR or elsewhere in Eastern Europe.

There is some evidence from the scholarly literature that local relations between mechanical engineering firms and their final customers in the automobile industry and elsewhere were damaged by the economic crisis. This is especially apparent in the car manufacturing industry. There is also evidence that supply relationships in the Baden-Württemberg machine industry are also becoming increasingly national or international in scope.[29] Semlinger (1995), for example, finds that firms which have traditionally concentrated production in-house in the vehicle and machine producing industries are beginning to turn to outside suppliers, but that these suppliers may be elsewhere in the country, or even abroad. Gary Herrigel (1996b) sees this as being bound up in wider processes whereby the kinds of flexible production found in industrial districts are being challenged by less locally embedded kinds of flexibility and an accompanying "de-regionalisation and re-regionalisation" process that will likely involve the "killing off" of the old regional order of production.

3.7 CAN RATIONAL CHOICE EXPLAIN TRUST AND COOPERATION IN INDUSTRIAL DISTRICTS?

I have already discussed the very considerable skepticism concerning rational choice explanations of human behavior among scholars studying industrial districts. In Chapters 4 and 5, I seek to show that rational choice can indeed provide us with important insights into trust and cooperation in these districts. Before doing so, however, I wish to

[29] Although Eastern European suppliers may not be as attractive to Baden-Württemberg machine manufacturers as some local commentators fear, several of the firms interviewed had tried Eastern European suppliers, especially in the Czech Republic. Almost invariably they had found that these firms could not produce to the required quality. Thus, Stuttgart machine tool makers have not been as likely to create international supply chains as other manufacturers (on whom see Thelen and van Wijnbergen 2003).

carry out a far more modest task, establishing whether interest-based motivations provide a good baseline explanation of cooperation in the two cases. There is already a lively debate on the causes of cooperation among scholars of industrial districts. Both communitarian and identity-based accounts suggest that trust and cooperation are not based on stable forms of material self-interest, while a small number of scholars argue to the contrary. In the remainder of this chapter I wish to make a case that trust and cooperation *are* largely based on stable forms of self-interest and that the encapsulated interest account provides an excellent account of how they work.

Skeptics of self-interest, including communitarian scholars such as Patrizio Bianchi and Giuseppina Gualteri (1990), argue that individuals' propensities to trust or to distrust others within industrial districts are the product of long-lasting historical structures of identity. Strong local collective identities foster trust and cooperation because they accentuate the collective good above that of the individual. Similarly, Frank Pyke and Werner Sengenberger (1992) argue that the kinds of cooperation to be found in industrial districts are made possible by the existence of communities with supportive sets of values. While Pyke and Sengenberger argue that self-interest does not disappear in this context, they claim that it is transformed so that it includes the welfare of others, as well as one's own future welfare. Individual gain is subordinated to a system of strongly held local communal values, which lay a stress on collective rather than individual advantage.

Other skeptics agree that identity trumps stable interests, but they argue against communitarian claims that identities are a product of deep-rooted historical communities. Charles Sabel, Jonathan Zeitlin, and their colleagues agree that the "co-operation, trust and community" that have "been crucial to the economic success of industrial districts such as those of the Third Italy" are "distant from prevailing conceptions of economic rationality" (Zeitlin 1992, 286). However, their claims about the sources of cooperation, trust, and community are quite different. They posit that identity is fundamentally malleable and reflexive, so that it is always open to reinterpretation. Thus, Charles Sabel (1993) argues that commonly agreed "stories," which circulate in industrial district, are the basis for trust. These stories arise when persons or groups agree on a common history that resolves their differences or renders them irrelevant.

In Sabel's description, "instead of expressing a consensus, the stories in this view are part of the process of creating it," (1993, 117) so that conflict, and "an almost mythologically extolled consensus coexist in high trust systems" (119). In short, the stories that underlie trust and cooperation are themselves the product of social construction, and may be revised. Old stories of distrust and noncooperation can be transformed into new stories in which people have "always" cooperated (Sabel 1993; see also Herrigel 1993). Recent versions of this argument suggest that the forms of collaboration found in industrial districts can, and must be, stretched to cover wider geographical spaces than those typical of industrial districts (Sabel 2004; Herrigel 2004).

Neither of these accounts entirely denies that self-interest, as economists and rational choice social scientists conceive of it, is relevant to behavior in industrial districts. However, they both dismiss it as being unrelated to the kinds of trust and cooperation that make these districts interesting. In contrast, the encapsulated interest account would focus on clashes and consonances of interest as being the key factor determining whether actors trust and cooperate with each other.

Most obviously, in a market context, we may expect that the degree to which actors' interests are consonant or divergent from each other will depend to a very great extent on whether or not they are in direct competition with each other. Actors that are direct competitors are likely to have a far narrower range of issues on which they may cooperate than actors that either make complementary products or actors that are engaged in a different phase of the production process, upstream or downstream. In situations where the success of one firm is likely to come at the expense of its competitors, we may expect that the firm and its competitors will be unlikely to share common interests, except over a narrow range of issues (such as those touching on the general health of the economic sector in question).

Which of these three accounts – communitarian identity, fluid identity, or stable self-interest – best explains the kinds of cooperation observed in the two cases? There is evidence from interviews that both personal relationships and community ties had some implications for trust and cooperation. One Italian interviewee spoke of how he could rely on orders from friends during difficult business periods and on the importance of local bonds, describing how his community was characterized by a "certain spirit of being able to help. There is not a closedness. I am among my fellow-parishioners, am inside my garden."[30] Another interviewee described "affections that you find in certain ways … of thinking, of sharing things. There is a lot here; there is a feeling, all in all. For me, I can think of phoning up: "I have a problem, can you help me?"[31]

However, the consequences of these ties were limited in scope and extent. The business actor who spoke of the importance of "affections" stressed that cooperative relations were impossible with direct competitors, as everyone wanted to sell the same thing.[32] Several other interviewees in Italy and Germany made it clear that they did not see any spirit of mutual cooperation or help in the area or the industry; firms could only be expected to cooperate when it was in their interests.[33]

For the majority of business actors who were interviewed, personal relations across firms seemed, in Andrew Lorenz's (1988) phrase, to be between "neither friends nor strangers." Like Lorenz's French engineering firms, interviewees often stressed the importance of personal contact and sometimes described their relationships using the language of friendship. But this does not usually seem to have involved deep emotional or ethical commitments to the other; rather, the language of friendship provided a compass to orient social relations that were primarily instrumental. This widespread set of expectations was well represented by one interviewee, who, after describing how his firm cooperated with others on the basis of friendly acquaintance, felt it necessary to add that "of course, this is a business world, and therefore you can't expect anything from the others, and the others can't expect anything from us."[34]

When describing their relations with third-party firms, interviewees repeatedly referred to the limits of the cooperation that might be expected from others or that they themselves would give. Indeed, they

[30] Firm Interview 14.

[31] Ibid., 16.

[32] Ibid. The second firm interview cited above (Firm Interview 14) did speak of the possibility of cooperation between competitors, but this cooperation was extremely limited in extent.

[33] Ibid., 13, also 1, 2, 7.

[34] Ibid., 2.

often seemed to have difficulty in understanding the notion that others might cooperate where it was not in their direct self-interest to do so. Some interviewees stated that their businesses had helped other firms in various ways in the past, and that they could imagine that other firms would help them in turn. However, in both Bologna and Stuttgart, interviewees seemed to believe that such help was unlikely without an immediate and tangible return, such as shares or indeed the takeover of the company in trouble.

In the words of an expert at a Bolognese economic research institute, who was familiar with the mechanical engineering industries of both Bologna and Stuttgart,

> I do not share the opinion of theorists of industrial districts that there is an almost fraternal trust [within these districts]. It is a trust in commercial terms, as developed among Italian entrepreneurs and German entrepreneurs. It is a trust that is based on relationships that become consolidated over time, and that are profitable for both parties.[35]

Thus, neither the packaging machinery cluster in Bologna nor the machine tool producing district in Stuttgart appeared to resemble the bucolic communities of trust depicted by Bianchi and others. While relationships of trust were clearly important, they were based on self-interest and perceptions of self-interest. There seemed to be little role for community spirit of the sort that Bianchi and other theorists have described. Personal ties and friendships did have some consequences for trust, but they always stood in danger of being trumped by economic interest.

In contradistinction to accounts such as Sabel's and Herrigel's that emphasize the fluidity of identity (and of perceived interests), actors' perceived goals seemed to approximate closely to those that might have been expected given their material interests. There was no evidence that trust and cooperation could be traced back to mutable collective stories defining actors' identities as Sabel and others argue. Business actors in both cases were indeed aware of the history of how industry had developed in their region and viewed it as a source of pride. However, there was no evidence of any connection between these stories and actual patterns of trust and cooperation among firms. While

[35] Interview with Silvano Bertini.

trust and cooperation played an important part in the organization of production, especially in the Italian case, the kind of "mythically extolled consensus" that Sabel (1993) posits as the key source of trust in industrial districts did not seem to exist.

Where trust and cooperation did occur between business actors, not only did they seem to be based in perceptions of economic interest but they seemed to follow predictable patterns. As the encapsulated interest account would predict, extensive forms of trust and cooperation were far more rare between direct competitors than between noncompetitors. Although firms frequently reported collaboration with other firms that offered complementary or noncompeting products, interviewees typically suggested that cooperation with competitors was difficult or impossible.[36] Firms frequently swapped information on a basis of diffuse reciprocity with other firms that offered complementary products. However, they did not do so with competitors. Similarly, interviewees (with one notable exception) usually reacted with incredulity or incomprehension to the suggestion that they collaborate over technical problems with competitor firms. In the words of one German interviewee:

> Cooperation in the sense of common development or a transfer of knowledge doesn't come into the question. One must protect one's know-how.[37]

This general reluctance to cooperate stemmed from competitive pressures. In the description of one Bolognese interviewee:

> In Bologna, there are probably another five firms which do this. If I have a problem to resolve in that the current machine produced by [my firm] does 60 packages a minute, and a customer wants a machine which can do eighty a minute, I have two possible routes. Change the machine completely, and make a new one, or try to modify the existing machine, to see if it can do 70 or 80 a minute ... I am sure that the other four or five firms also have had this problem ... there is competition, but there is not exchange of information. Absolutely not. Because this is part of power in the end, leads to customers and competitiveness. This is clear. Because if there are customers who [need] eighty packagings a minute, if I do not give a positive response, but the competitor gives a positive response, I will lose the customer.[38]

[36] Firm Interviews 1, 3, 7,8, 12, 18, 20, 25, 26, 27, 32.

[37] Ibid., 20.

[38] Ibid., 3. A similar perspective was expressed in Firm Interview 1.

These findings are not surprising. They confirm what most scholars of political economy would expect, that the likelihood of cooperation among firms is in large part a product of relatively stable material interests. They do, however, make it clear that neither communitarian nor identity-based accounts provide a sufficient basis for explaining observed outcomes, contrary to the arguments of many scholars of industrial districts. This does not mean that these accounts are worthless – for example, one may easily see how actors' identities may play an important role (and perhaps be shifted through persuasive argument) in contexts where their interests are unclear. But it does suggest that their explanatory purchase is limited. By demonstrating the manifest importance of interests to trust and cooperation, I hope (1) to show that the evidence contradicts accounts that deny that material interests play an important role in explaining trust and cooperation in industrial districts, and (2) to clear the ground for a more sophisticated account of precisely *which* factors influence actors' interests and expectations, and hence their propensities to trust or distrust each other.

Thus, in the succeeding chapters, I hope to build on the work of a small group of scholars, most prominently including Gabi dei Ottati (1991; 1994; 1996a) and Edward Lorenz (1988; 1992a; 1992b), who *do* argue that self-interest plays an important role in explaining cooperation in industrial districts. Both of these scholars build upon Michael Taylor's work (1976; 1982; 1987) on collective action and many-sided relationships to argue that community can explain the kinds of trust and cooperation to be found in industrial districts. In a community where there are thick, many-sided relationships among actors who know that they are likely to continue to deal with each other into the indefinite future, these actors may have a good, self-interested reason to behave in a trustworthy manner. They know that if they do not, others will have many future opportunities to punish them for their bad faith. By the same logic, these others will have good reason to trust them across a wide variety of matters.

This provides a very good start for a rational choice account of trust and cooperation in industrial districts, but it still leaves much to be explained. By relying on Taylor's game theoretic account, dei Ottati and Lorenz effectively invoke folk theorem results that state that cooperation is *possible* in contexts of repeated interactions within appropriate parameters, but which cannot predict whether cooperation is more or less likely within different communities.[39] In short, they lack a theory of equilibrium selection – even within communities, many different cooperative equilibria are possible, ranging from a complete lack of cooperation, through various intermediate stages of partial or limited cooperation, to extensive cooperation across all relevant issue areas.[40]

In order to provide more specific predictions, we need a better account of the mechanism or mechanisms that lead actors to converge on one equilibrium rather than another. As I have argued in Chapter 2, institutions, properly considered, provide just such a mechanism. By providing actors with information about how other actors belonging to general classes may be expected to behave in certain situations, and about how others will expect the actors themselves to behave, institutions will guide actors toward certain equilibria and away from others. I turn to the task of demonstrating the importance of institutions (and institutional variation) in explaining trust in the next chapter.

[39] Note, however, that Lorenz (1999) has modified his position over time, and he criticizes game theoretic approaches to trust. Note also that Michael Taylor has changed over time from a vigorous proponent of rational choice theory to a harsh critic; see Taylor 2006.

[40] Lorenz (1999) discusses "procedural rules" of gradual confidence-building that make trust more likely, but it is unclear that these provide a theory of equilibrium selection of the kind that I discuss here.

4

Trust and Institutions in Industrial Districts

4.1 INTRODUCTION

Rational choice accounts of the sources of trust provide a strong basis for understanding the sources of cooperation in industrial districts. Here I inquire whether a rational choice account can explain not only broad patterns of cooperation across the two cases but also differences between them. The existing literature on industrial districts is dominated by detailed case studies of single industrial districts, with remarkably little comparative work. Although we know that industrial districts vary widely in their patterns of economic organization, we know very little about the sources of this variation.

Moreover, the kinds of variation that we see present a quite perplexing puzzle for the political culture perspective, which has been the dominant approach to the explanation of cross-national variation in trust and cooperation. Robert Putnam's (1993) arguments about social capital posit that "civic" factors have been key to the relative economic and cultural success of Northern Italy – and the stagnation of the South.[1] Putnam specifically refers to industrial districts as key

[1] Putnam denies that he is part of the political culture tradition. Sidney Tarrow (1996), however, disagrees, locating him firmly within it. For other important critiques of Putnam, see Bagnasco (1994; 1999), Levi (1996), Jackman and Miller (1996a; 1996b; 2004), Sabetti (1996), Trigilia (1999).

microlevel illustrations of the links between civics and economics that he claims to have identified.[2] In his argument:

> What is crucial about these small-firm industrial districts, conclude most observers, is mutual trust, social cooperation, and a well-developed sense of civic community – in short, the hallmarks of the civic community. It is no surprise to learn that these highly productive, small-scale industrial districts are concentrated in those very regions of north-central Italy that we have highlighted as centers of civic traditions, of the contemporary civic community, and of high-performance regional government. (Putnam 1993, 163)

However, even though Putnam's broader findings on the relationship between economic and cultural factors have been claimed as evidence in favor of political culture (Inglehart 1990; 1997), the existence of industrial districts poses a quite perplexing puzzle for the political culture account. Italy appears to have lower levels of generalized interpersonal trust than other European countries. Nonetheless, it has enjoyed considerable economic benefits from a form of production that involves extensive trust and cooperation among firms (Locke 1995; Sforzi 1996). Furthermore, as I will argue, "high trust" countries such as Germany seem to have considerable difficulty in replicating these cooperative forms. Nor can Italy's success be explained by the kinds of intracountry variation that Putnam's (1993) account highlights. Survey evidence indicates that the "civic" North of Italy (where the majority of industrial districts are located) does indeed have a higher level of diffuse interpersonal trust than the South, but that Northern Italy had lower levels of diffuse interpersonal trust than all other West European countries where data was collected during the 1980s (Inglehart 1990).

Thus, we might reasonably expect that a political culture account would predict that high-trust forms of cooperation would be more likely to occur in Germany, or other industrialized democracies where the level of diffuse interpersonal trust is relatively high, than in Italy, which political culture scholars have traditionally depicted as a paradigmatic example of low trust. Instead, we find that certain kinds of

[2] As Locke (1995) and Burroni and Trigilia (2001) point out, industrial districts are increasingly to be found in other parts of Italy than the North, suggesting that the stark division that Putnam (1993) and other political culture theorists draw between North and South is to some considerable extent misconceived (Tarrow 1996).

high-trust cooperation seem to be much more difficult to achieve in Germany than in Italy. Why would this be so?

In this chapter I argue that this puzzle – why high-trust forms of production appear to be easier to achieve in a country with a low level of diffuse interpersonal trust as measured by survey data than in a country with high levels of diffuse impersonal trust – can best be explained if we look at differences between institutions. It is by paying proper attention to institutions – and how they affect trust and cooperation – that we can best explain the observed variation. I argue that an institutionalist account of the kind I lay out in Chapter 2 provides a better understanding of cross-case variation than a political culture argument by tracing out the relationship between formal institutions, informal institutions, and trust.[3] It also provides an empirical illustration of exactly how institutions can affect trusting relations among actors. The rational choice literature on institutions and cooperation between actors parallels,[4] but has not been comprehensively applied to, the study of trust. A comparative analysis of the effects of institutions on trust would suggest that variation in institutions will be associated with variation in trust between actors in predictable ways. The two case studies of trust and cooperation between firms in Stuttgart and Bologna provide just the kind of variation that might be expected. In contrast, the predictions of the political culture model would appear to be falsified.

First, I discuss the different predictions that institutionalist and political culture approaches to Italy and Germany would make, examining differences in both political culture (as measured by the World Values Survey) and institutions between the two countries. Next, I examine the comparability of the two cases on one vital aspect of cooperation in the productive process – cooperation between final producers and subcontractors in the making of machines. Then

[3] This work should not be taken as an attack on the study of political culture *tout à fait*. I share with students of political culture the conviction that "cultural" variables play an important role in the explanation of political behavior. See Farrell and Knight (2003). Although the "oversocialized" account of human action proposed by some political culture theorists (e.g., Eckstein 1988), and embraced in Inglehart (1990), is highly problematic, other conceptions of political culture (e.g., Elkins and Simeon 1979; Barnes 1988) view it in a much less totalizing way.

[4] For example, Miller 1992; Greif 1994; North 1990; Milgrom, North, and Weingast 1990; Greif, Milgrom, and Weingast 1990; Calvert 1995a; 1995b.

I set out evidence from interviews about differences in cooperative practices between the two countries. Finally, I suggest that the differences between formal and informal institutions that are discussed in this chapter have important implications for debates over economic development.

4.2 POLITICAL CULTURE, INSTITUTIONALISM, AND CROSS-NATIONAL COMPARISONS

As Robert Jackman and Ross Miller (2004) observe, the political culture and institutionalist approaches to the understanding of trust and cooperation are based on starkly different understandings of the world and political motivations.[5] While the political culture approach tends to argue that people are motivated by learned orientations (Eckstein 1988; 1996; 1997), rational choice focuses more on how incentives affect individuals' intentional actions (Levi 1997). These differing theories may give rise to different predictions when learned orientations (insofar as we can determine them from the survey data that political culture scholars typically invoke) and institutional incentives seem to point in different directions.

Just such an empirical disagreement is at stake when trust and cooperation in Germany and Italy are compared. Indeed, Italy enjoys a quite particular status within political culture research, which would seem to argue strongly against it being a likely site for high-trust forms of production. The political culture approach to comparative politics has treated Italy as its paradigmatic example of how an advanced industrialized democracy may have an unhealthy political culture and be subject to political "fragmentation, isolation and alienation" (LaPalombara 1965), ever since Almond and Verba's *The Civic Culture* appeared in 1963.[6] Political culture accounts have typically depicted Italy as a prime example of low interpersonal trust (Almond and Verba 1963; LaPalombara 1965; Inglehart 1990). Survey data from the 1980s suggests that Italy has had lower levels of interpersonal trust among its citizens than any other European country (Inglehart

[5] See also Taylor 1989.

[6] On the uses and abuses of Italy as a paradigmatic example of political dysfunction, see Sabetti 2000.

1990); even though it has improved gradually over recent years, it is still lower than in most other democracies (Inglehart 1997).

In contrast, research in the political culture tradition argues that Germany enjoys relatively high levels of diffuse interpersonal trust (Inglehart 1999). Table 4.1 shows a simple tabulation of data from the World Values Survey comparing levels of diffuse interpersonal trust in Italy and West Germany in 1990, and in Italy and postunification Germany in 1999.[7]

As is immediately obvious, there are stark differences in the responses of Italians and Germans to the question of how much they trust their fellow nationals. In 1990, only 2.7% of Italians trusted other Italians completely, while 16.1% of West Germans did. In 1999, even fewer Italians – only 2.5% – trusted their fellow citizens completely, while 26.6% of Germans did. There are less striking, although still substantial differences in the percentage of citizens who at least trusted their fellow citizens "a little." In 1990, 35.6% of Italians trusted their fellow citizens a little, while 49.4% of West Germans did. In 1999,

TABLE 4.1. *Respondents' level of trust in fellow nationals in Italy and West Germany 1990, and in Italy and Germany (both former West and former East Germany) 1999 (percentage values).*

	Italy 1990	West Germany 1990	Italy 1999	Germany 1999
Trust completely	2.7	16.1	2.5	26.6
Trust a little	35.6	49.4	44.2	51.5
Neither trust nor distrust	40.7	15.4	39.0	15.1
Not very much	16.6	16.3	12.0	6.1
Not at all	4.4	2.8	2.3	0.7
Total	100% (1987)	100% (1905)	100% (1966)	100% (1967)

Source: World Values Survey (http://www.worldvaluessurvey.org/).

7 The data are drawn from answers to question G007: "Using the responses on this card, could you tell me how much you trust [people of your nationality] in general?"

the respective figures were 44.2% and 51.5%. In general, levels of diffuse interpersonal trust among Italians appear to be dramatically lower among Italians than among Germans; while 38.3% of Italians in 1990 and 46.7% of Italians in 1999 had at least some trust for their fellow citizens, the respective figures for Germans were 65.5% in 1990 and 78.1% in 1999. If political culture scholars' emphasis on the importance of diffuse interpersonal trust as an explanatory variable is correct, it is hard to escape the conclusion that Italians should be considerably less trusting of each other than Germans. We should therefore expect high-trust forms of cooperation to be more prevalent in Germany than in Italy.

An institutionalist perspective provides quite different predictions. Institutions are likely to affect business cooperation both through shaping the incentives that firms have to concentrate production in-house or to put work out to subcontractors and through directly shaping the ways in which business actors can or cannot trust each other. First, the factors stressed by traditional comparative political economy approaches – for example, national-level formal institutions – are likely to affect patterns of cooperation directly without necessarily affecting trust. Here, one might expect that labor law and industrial relations systems will play an especially important role in shaping firms' incentives either to carry out work within the firm or to put it out to others.[8] Under some systems of labor law and industrial relations, it makes sense for larger firms to put work out – by so doing, they can lower their own costs while delegating production to smaller firms that may carry a lower legal burden of responsibilities, be better able to evade the effect of cumbersome laws and/or unionization, or both. Under other systems, where small firms and large firms carry similar burdens, there will be less incentive to put work out, and more incentive to keep it in-house. Thus, entirely independent of institutional consequences for trust, one may expect different patterns of cooperation between firms in countries with different labor systems.

The existing literature suggest that there are indeed clear differences in the labor law and industrial relations systems of Italy and Germany. Italian labor law imposes a much lighter effective burden

[8] I am grateful to an anonymous reviewer for suggesting that I incorporate discussion of labor relations.

on small firms than on large ones (Regini 1997), providing stronger incentives toward the disintegration of production than in countries such as Germany, where the differences between the obligations of small firms and large ones are not as marked.[9] These macrolevel institutional differences will plausibly affect cooperation between firms indirectly, through providing a greater incentive toward subcontracting out work than exists in countries like Germany, hence making Italian firms more likely to rely on cooperative relationships with other firms where these are available.

Similarly, there are important differences in industrial relations systems between the two countries. The Italian national industrial relations system only weakly constrains firms and unions (Regini 1997). As described by Alacevich et al.:

> The Italian system of industrial relations is characterized by a low level of institutionalization. ... Rules, procedures and institutions are not regulated on the formal level, by law. ... Collective bargaining, and the practices of concertation and consultation have thus traditionally been developed in a way that is mostly voluntaristic and informal. (2000, 11, my translation)

As a result, while national agreements provide a broad nonbinding framework for specific deals, the key level of decision making for most issues is the "territorial" or local one, where local unions bargain directly with firms.[10]

Germany, in contrast, has a much stronger bipartite national and state-level industrial relations system, albeit one that has come under increasing challenge in recent years (Thelen and van Wijnbergen 2003; Silvia and Schroeder 2007). Here, deals struck between *Gesamtmetall* (the metal and mechanical engineering industry employers' association) and *IG-Metall* (the metalworkers' union) have traditionally set a broader precedent for wage bargaining in the country as a whole. This

[9] Arnaldo Bagnasco describes how these rules helped precipitate a shift in production to small firms: "With the increase in job security and labor costs, industrial firms reacted with a strategy that has been described as the decentralization of production. Over the course of the 1970s, many processes that had previously been centralized within large or medium sized manufacturing firms came to be put out to small firms, where labor rules could more easily be circumvented, and where it was easier to hire and fire as the economic cycle dictated, and where in general labor costs were thus lower" (1988, 27).

[10] Interview with FIOM official.

system has come under challenge because of the weakness of employers' associations in which large firms (which have an interest in continued industrial peace) predominate, and the resulting defection of many small firms, which do not see their interests adequately reflected (Thelen and Kume 2006). The result is a greatly increased emphasis on firm- level bargains that may supplement or modify the application of national agreements. While the German national labor relations system remains far stronger than the Italian system, it is vulnerable to continued defections over the longer run.

These differences help explain variation in incentives to outsource production between the two national systems. In systems such as Italy, where small firms have an easier time under the law and may more easily evade those regulations that do apply to them (and perhaps efforts at union organization), we may expect that large firms will have greater incentive to outsource production and to shrink their internal operations than in countries such as Germany. Indeed, Italian industrial districts received much of their initial impetus from the introduction of stricter labor laws in the wake of the *autunno caldo*. Germany suffered no similar shock.

If these macrolevel institutional differences were the end of the story, then there would be little reason to believe that differences in trust helped explain cooperative outcomes. However, even if the standard political economy approach can explain gross differences between Italy and Germany, it cannot explain the high degree of variation in patterns of cooperation that one may observe *within* the Italian national system. Even though this national system has a common set of laws and national industrial relations rules, we observe vast differences in patterns of cooperation in different regions and localities, even across apparently similar sectors of production. The existing literature is emphatic (Bagnasco 1977; Trigilia 1986) in stressing that localized forms of production in Italy are not a simple product of large firms' outsourcing strategies, and that local and regional variation is the result of differences between semiautonomous local economic systems. Models of production vary along with these local systems, from the conventional large- firm production model that traditionally characterized much of the northwest, through different forms of networked firm and small-firm network economies in the Veneto, Emilia-Romagna, and Tuscany, to the various models of small-firm production (some inefficient, some beginning to achieve economies of cooperation) in the South (Burroni and Trigilia 2001). These different models of production are associated with stark differences in subcontracting practices, ranging from models based around traditional large firms that organize most of their production in-house (most prevalent in the northwest), through models involving extensive collaboration among small firms (the classic industrial district model) to small-firm economies involving little cross-firm collaboration.

In the German system, in contrast, differences in forms of production across regions are far more muted, with the obvious exception of the stark divergences between former West Germany and former East Germany (which stem from several decades spent under very different institutional systems). Although Gary Herrigel (1996) claims to identify different "regional orders" within Germany, other scholars (Voelzkow 1999; Voelzkow and Glassmann 2001) find little evidence that these regional orders exist, and they argue that Germany is better understood as possessing a single national model of production.

In order to explain why Italy is composed of a congeries of local economies (Locke 1995), while Germany has far less internal variation, institutionalists need to turn to a second set of factors – the interaction between formal national institutions and informal local ones. Much recent work in political economy highlights the importance of institutions capable of underpinning impersonal transactions as a component of economic success (North 1990; Greif 1994). Here, again, there are important differences between Germany, where the state has mostly succeeded in creating laws, regulations, and forms of delegated authority that underpin such transactions, and Italy, where the state has at best had modest success in creating and implementing such institutions (Bagnasco 1994).

In Italy, as described by Marino Regini:

> even when public policies apparently assign a leading role to state regulation, ... mechanisms for circumventing them are often set in motion; or else the state rules are only weakly and inefficiently implemented, with the consequence that even the opposite result may be achieved. (Regini 1997, 106)

Not only is the Italian state incapable of properly enforcing institutional rules, but the court system is cumbersome, inefficient, and perceived as being partial and politicized (Volcansek 1990). This has had important consequences for the political economy of Italy, leading to a high level of insecurity in the business environment (Trigilia 1996; Arrighetti, Bachmann, and Deakin 1997).

However, the weakness of national formal institutions creates opportunities for actors at the local level, which has allowed for the creation and persistence of local economies that are highly reliant on informal rules (Regini 1997; Locke 1995; Bagnasco and Trigilia 1993). It is important to note that localized systems of informal rules are by no means necessarily efficient and do not come into being "in order" to support cooperation. Indeed, the localized informal economies that have emerged in many parts of Italy are highly dysfunctional and reflect the differing distributional interests of more or less powerful actors rather than any functional imperative. Even so, in certain of these economies, complex informal rules have come into being that do support certain kinds of cooperation, supplementing, and, to some extent, displacing the formal laws and regulations of an inefficient state. The existing literature suggests that industrial districts typically involve just such rules (Trigilia 1989; Bellandi 1996; Bianchi 1993; Brusco 1990). As described by Sebastiano Brusco, in industrial districts:

> alongside state regulations, there is a second set of rules that derives from the community to which all the companies belong. This set of rules, shared by everyone and to which everyone has to adapt, originates in civil society, and also carries a series of sanctions: whoever breaks the rules of the game is excluded from the community and can no longer work with it. (Brusco 1990, 182)

Although Brusco specifically disavows rationalist interpretations of industrial districts, his description of these rules is compatible in every important respect with Jack Knight's (1992) definition of an informal institution:

> Social institutions are sets of rules that structure social interactions in particular ways. These rules (1) provide information about how people are expected to act in particular situations, (2) can be recognized by those who are members of the relevant group as the rules to which others conform in these situations, and (3) structure the strategic choices of actors in such a way as to produce equilibrium outcomes.[11]

[11] Knight 1992, 54. Brusco emphasizes the importance of local tradition in creating an appropriate context for modern cooperation. Carlo Trigilia (1986) similarly

Thus, in short, Italy presents an example of an economy where deficiencies in the enforcement of formal institutions have created space for a profusion of local economies to come into existence, each based on specific informal rules applying to a local community of actors.[12]

The prevailing literature suggests that Germany presents no such complicated mixture of weak formal institutions and strong local informal institutions. While local identity is still politically important in some contexts, the existing literature suggests that formal institutions are credible and effective. As Wolfgang Streeck puts it, "the institutions which embed [Germany's] economy and shape its performance are politically negotiated and typically legally constitutionalized, rather than commanding compliance as a matter of informal obligation."[13] The result is a system in which there is little evidence of the kinds of local "community" rules that play such an

stresses the importance of local "sub-cultures" in paving the way for industrial districts. Cultural factors of this sort may indeed play an important role in creating a context where actors realize that certain forms of cooperation are possible and beneficial. Symbols and cultural beliefs may disclose or foreclose possibilities of action within a specific community and thus affect actors' behavior and the institutions that arise from these interactions. Here, their role is anterior to the power relationships that I emphasize in Chapter 5; they may partially define the range of "possible" equilibria over which actors may bargain. For further development of these arguments, see the discussion in Ensminger and Knight (1997), Johnson (1997) and Farrell (2003). I am grateful to Carlo Trigilia for reminding me of the importance of these factors.

12 The case study of the Bolognese packaging industry should *not* be taken as a proxy for the Italian economy as a whole. While I examine the effects of national-level institutions on this industrial district, these institutions permit a great deal of variation. In contrast, the degree of national homogeneity in Germany is arguably far higher (Voelzkow 1999; Glassmann and Voelzkow 2001). One may further note that there is an important difference between the two perspectives that I am testing here. The political culture approach has usually been applied at the national level, perhaps because of its heavy reliance on national-level survey data. I am unaware of any political culture studies that examine the phenomena of interest for this book at the subnational level, apart from Putnam (1993) and Inglehart (1997), which I discuss in the main text. Both of these seek to draw a strong distinction between North and South Italy – but as I note, political culture scholars claim that even North Italy has lower trust levels than Germany or any other West European country. Furthermore, political culture scholarship (Putnam 1993) has sought to explain the particular phenomenon at hand – trust and cooperation in industrial districts – as a manifestation of the "civicness" to be found in Northern Italy as a whole. Thus, it is entirely reasonable to claim on the basis of the existing literature that the political culture approach would predict lower levels of trust in the Italian case than the German one.

13 Streeck 1997, 37; see also Arrighetti, Bachmann, and Deakin 1997.

important role in Italy (Voelzkow 1999; Glassmann and Voelzkow 2001). Instead, national level formal institutions appear to predominate. Furthermore, these institutions appear to be relatively predictable. Arrighetti, Bachmann, and Deakin (1997) find on the basis of survey research among firms that the stability and predictability of the German court system allows firms to calculate the costs and consequences of legal action, making "*actual* recourse to law unnecessary except in very unusual cases" (or, in game theoretic parlance, off the actual path of play).[14]

What effects are these institutional differences likely to have on trust relations? As I have argued in Chapter 2, formal and informal institutions are likely to be associated with quite different outcomes in terms of trust among individuals. Formal institutions involve written rules that are typically enforced by a third party such as the state; informal institutions involve unwritten rules that are typically enforced through bilateral relationships and multilateral information sharing within a given community of actors. Formal institutions, such as laws or legally enforceable contracts, are usually relatively specific; they thus may induce clear expectations about actors' likely strategies under circumstances that are foreseen and addressed by the institution. By the same token, they are likely to provide weak guidance when unanticipated circumstances arise. Informal institutions, in contrast, are diffuse unwritten understandings; while they may provide less precise expectations about actors' strategies, they are more easily adapted to previously unforeseen contingencies.

Thus, in environments dominated by formal institutions, we are likely to see a high degree of *confidence* over issues that are directly and explicitly covered by the institutions in question. In a very important sense, trust will be redundant over these issues, as actors will be able to rely on the clear expectation that these institutions will be enforced in the case of default. In contrast, environments dominated by informal institutions will see a high reliance on *trust*. Institutions will be important less because they spell out a clear and determinate set of expectations in advance, than because they provide a means of ascertaining how different classes of actors are likely to behave in a given situation that was not precisely covered by the institutional rule.

[14] Arrighetti, Bachmann, and Deakin 1997, 181.

Even though environments dominated by informal institutions are not necessarily high-trust environments (see my subsequent discussion of the institutions governing the Sicilian Mafia), they are environments in which trust (as opposed to confidence in impersonal institutions) is important to whatever forms of exchange do take place.

Furthermore, when *appropriate* informal institutions are in place (i.e., institutions that suggest actors should cooperate over a wide range of issues) we may expect high-trust forms of cooperation. Appropriate institutions, in this context, are institutions that work in at least one of two possible ways. First, they may provide actors with grounded reasons to expect that actors who they know belong to certain general classes (i.e., they know that *b* is a member of class *B*) will behave in a trustworthy fashion with regard to matters of a particular kind. Second, they may support forms of communication that (1) allow actors better to assess the reputation of particular others so as to assay these others' trustworthiness and (2) give these others incentives to behave in trustworthy ways (because they know that if they do not, they will be likely to acquire a bad reputation and be shunned).

Formal and informal institutions will have different consequences for the ability of actors to make credible commitments to each other, and thus for trust. Actors who wish to make credible commitments to each other in a setting where there are appropriate informal institutions will be able to do so with regard to a broader set of matters, some of which may not be readily definable in advance (i.e., an unanticipated problem may arise after actors have already committed to each other). In contrast, actors in a setting that is dominated by formal institutions will only be able to use these institutions to anchor credible commitments over a narrower set of issues, and they will have considerable difficulty in making institution-based commitments regarding matters that cannot be anticipated in advance. This is not to deny that actors in formal institutional settings can make informal arrangements with each other – they can and frequently will (Macaulay 1963, MacNeil 1978, Williamson 1985). It is to say that these informal relationships will be over a narrower range of issues and will not be able to invoke the same kinds of collective punishment as are possible under many informal institutional arrangements (Dixit 2004).

Thus, the institutionalist perspective advanced in Chapter 2 would lead one to predict quite different outcomes in the Italian and German cases. Machine producers in Stuttgart are embedded in a relatively homogenous set of national-level formal institutions (Voelzkow 1999; Glassmann and Voelzkow 2001). Given the predominance of formal institutions in Germany, one would expect these producers to have a relatively high degree of confidence based on formal institutions, but also to have relatively low levels of trust beyond those kinds of trust supported by purely personal relationships, in the absence of binding and appropriate informal institutions to support this trust.

Machine producers in Bologna, in contrast, operate in an environment where formal institutions are much weaker and less predictable in their effects. The existing literature on Italian industrial districts (Trigilia 1989; Bellandi 1996; Bianchi 1993; Brusco 1990) argues that these local economies are characterized by strong informal rules that guide actors belonging to the community in question to cooperate. Thus, one would expect these actors to have relatively low degrees of confidence based on predictable formal institutions but high degrees of trust across important issue areas.

So, the political culture approach and the institutionalist approach set out in Chapter 2 would lead one to make quite different predictions about the consequences of trust for economic production in the two case studies. Political culture arguments claim that diffuse interpersonal trust, as measured by survey data, is the key variable underlying economic cooperation and thus economic success or failure (Inglehart 1990). Given that diffuse interpersonal trust is far higher in Germany than in Italy, this would lead one to predict that high-trust forms of production will be characteristic of Germany rather than Italy. Thus, a political culture perspective would lead one to expect to find greater trust among producers in the German than in the Italian case study.

An institutionalist perspective, in contrast, would lead one to focus on two factors. First are the consequences of macrolevel institutional differences (e.g., differences in the institutions governing labor relations) for patterns of cooperation. *Ceteris paribus*, we may expect that larger firms will be more likely to subcontract out production in countries where labor laws and labor relations give cost advantages to small producers, independent of expectations over trustworthiness. Given differences in the relevant labor institutions, this would lead one to expect that there will be more resort to subcontracting in Italy than in Germany. Second, and far more central to my account, are the differing implications of formal and informal institutions for trust. In Italy, the relative weakness of formal institutions at the national level is associated with a highly variegated economy, in which localized systems of community rules play an important part, especially in industrial districts. In Germany, strong formal institutions are associated with a relatively uniform political economy, in which there is little evidence of the sorts of strong, localized informal institutions that structure economic relationships in Italy. A national system such as Germany, in which formal institutions predominate will ipso facto be less likely to support trust-based cooperation and more likely to support relations of confidence over cooperative relations that are defined through contracts or other formal arrangements. A national system like Italy, in which formal institutions are relatively weak, and localized sets of informal institutions play an important role, will make actors more likely to rely extensively on trust. In local economies where there are appropriate rules, such as industrial districts, we may expect relatively extensive forms of cooperation based on trust among members of the relevant community. Thus, an institutionalist perspective would lead one to expect more extensive trust (and trust-based cooperation) among business actors in the Italian than in the German case.

4.3 COMPARING TRUST AND COOPERATION ACROSS CASES

By studying machine producers in Stuttgart and Bologna as comparable cases (Lijphart 1975), we can assess the relative merits of political culture and institutionalist accounts in explaining patterns of trust and cooperation among producers in industrial districts. As discussed in Section 5.3, Italy and Germany provide a substantial degree of variation on the relevant independent variables for the two accounts – diffuse interpersonal trust (for the political culture account) and institutional form (for the institutionalist account). By the same token, variation on key extraneous variables is minimized. Both case studies involve producers in the mechanical engineering industry, which face broadly similar problems across many aspects of the production process. In both case studies, there are dense networks of interaction between these firms, which are mediated by local organizations (the VDI and VDMA, in the German case, and the *Istituto Alderi-Valeriani*, a technical training institute, in the Italian case).[15] While differences in network structure (Wellman and Berkowitz 1988; White 1988; Kogut, Shan, and Walker 1993; Padgett 1993; Locke 1995) may have consequences for trust and cooperation that cannot be captured in a study of this sort, interview data suggests that there is at least a rough similarity between the gross network characteristics of the two case studies.[16]

As traditional political economy approaches predict, there are important differences in the industrial structures of the two cases. Mechanical engineering firms in Stuttgart frequently have long-standing sales relationships either with car manufacturers or their main subcontractors, which are often based locally. In contrast, Italian packaging machinery firms tend to seek their customers in international markets; the Italian market for these machines is relatively small.

More pertinently, there is strong evidence that some differences in structure do indeed stem from differences in labor market institutions. As discussed in Chapter 3, labor unrest in the late 1960s and early 1970s in Italy, followed by subsequent legal reforms, gave smaller firms a considerable competitive advantage, as they were not subject to some of the more onerous legal requirements (and found it easier in practice to evade these requirements when they were so subject).[17] The result is a far greater relative number of very small firms in Italy than in Germany. However, these macrolevel differences in industrial structure by no means are sufficient on their own to explain observed

[15] For more detailed discussion of the organizational settings of these districts, see Farrell and Holten 2004 and Glassmann 2004.

[16] For an exemplary study of how the shape of networks may affect power relations in industrial districts, see Locke 1995. To the extent that informal institutions may help shape network characteristics (by making individuals more likely to transact in specific ways with other actors), some of the factors that Locke identifies will be captured in my argument. But many will not. I know of no very good way of reconciling the insights of these two approaches; the network approach (at least as it is presented by Wellman and Berkowitz (1988) and other prominent network theorists) rests on a quite different social ontology than does the rational choice paradigm. For an interesting, if not entirely satisfactory, attempt at reconciliation, see Markovsky 1997. For a much less ambitious argument about the interplay of network characteristics and coordination games, see Farrell and Drezner (2008).

[17] For discussion of change in Italian labor law over the last 35 years, see Regini 1983; 1997; Regini and Regalia 1997; 2004; and Locke 1995.

differences in patterns of cooperation. In order to explain variation within Italy, it is necessary to turn to more locally specific factors. There is a high degree of regional and local variation in small firm patterns of production in Italy. Indeed, not only are national labor market institutions insufficient to explain the wide differences between localized economies based around shared sets of informal rules, but the causal arrow may run in the other direction. The importance of local sets of rules helps explain how labor market regulation actually works, and why national industrial relations agreements are so weak. As Marino Regini (1997) and Ida Regalia (Regini and Regalia 1997) describe it, labor market concertation has played a far more prominent role at the micro- than at the macrolevel in Italy, with strongly localized forms of "microconcertation" emerging in some localities, most notably in industrial districts, because in these districts:

> flexibility and cooperation were so deeply embedded in the social fabric and so strongly sustained by community trust relationships that they often seemed to be naturally part of the employment relationships, and therefore in no need of institutional mediation by the representative institutions. (Regini 1997, 112)

As described by regional union officials in Emilia-Romagna, the key role of the local level in determining industrial relations outcomes stems directly from the high degree of variation among local and regional economies in Italy:

> The national contract imposes rules on the territory, but the national contracts are not laws. ... A rule laid out in the national contract, may be applied or not applied in the territory, made better or made worse. Thus, they are in a feudal structure. It seems to me that many of our union [structures] are linked to how one works, how one organizes industry in the different parts of Italy. Big industry is in Turin, with Fiat – one can see others, but the model is Fiat. Here in Emilia, it is very different – there are many small firms, and it is a different model in social and historical terms. In the Veneto, it is different again. Thus, I think that it would be impossible, it would be difficult to have a national level that could take account of these differences.[18]

Thus, not only can one not simply explain differences in patterns of cooperation as a by-product of industrial relations institutions but

[18] Interview with FIOM representative.

Italian macrolevel industrial relations structures, and their differences from structures in other European countries, are likely a result of differences in the strength and importance of localized patterns of production.

Institutional differences also have more particular consequences for trust, and hence for cooperation, as may be seen if we compare how Italian and German firms deal with an extremely important part of the production process. Mechanical engineering firms in Italy and Germany face a fundamentally similar problem – the decision over whether to produce important components of a machine in-house or to put it out to specialized subcontractors. This problem touches on key issues of trust and cooperation, in that it precisely involves cooperative situations where all contingencies (and possible reactions to those contingencies) cannot be fully anticipated. We can therefore expect variation in actors' responses (their willingness to entrust key parts of the production process to outside firms) to be substantially affected by their beliefs over whether or not they can trust these others to behave appropriately in such contingencies.

These problems arise because both the packaging machinery and machine tool industries involve the production of specialized machinery to order, typically for large customers, with a substantial amount of bespoke modification to the machine according to the customer's particular needs. Thus, firms in both these branches of the mechanical engineering industry are subject to significant uncertainty over what will be required to produce a particular machine. A substantial literature in economic sociology addresses this "flexibly specialized" form of production and the forms of organization that it involves (Piore and Sabel 1984; Hirst and Zeitlin 1992).

Flexible specialization imposes particular requirements on the vertical organization of production. Firms face difficulties stemming in large part from three factors – variability in demand, the frequent need to change the specifications for orders on the fly, and the need for high-quality components compatible with the rest of the machine. Firms in both sectors experience a high degree of variability in demand – lulls in work may be succeeded by rapid spikes of activity in a quite unpredictable manner. Furthermore, producing specialized or customized machinery often requires that firms change specifications on the fly to meet the changing needs of customers. In theory, it makes sense for firms to subcontract out the manufacture of components where possible: this provides them with a higher degree of flexibility while allowing them to avoid paying for expensive machinery that would lie idle much of the time. However, if final producers are to rely on subcontractors to produce specialized components for these machines, they require that these subcontractors be themselves highly flexible, able to respond rapidly to changing needs, and to work for long hours on short notice. Further, it is important that each subcontractor produce well-tooled components compatible with the other components of the machine, which may be produced by other firms. The buyer firm needs to engage in a process of complex coordination, which is only possible if it is sure that each subcontractor is preparing components to high standards of quality. Again, this requires that firms be highly responsive to the technical needs of the buyer firm and prepared to make changes in the course of the production process.

This involves commitment problems for both parties. The firm putting out work has to be sure that it can rely on the subcontractor to respond flexibly, and with a high degree of attention to technical detail, at short notice. However, this sort of flexibility involves short-term costs for the subcontractor, which may find itself unexpectedly having to commit its staff to work overtime, or over weekends and holidays, so as to meet a particularly urgent order from the buyer firm. In the absence of some substantial arrangement between the two parties, it will often not be in the interests of the subcontractor to respond to the needs of the ordering firm; working over a weekend or during the holiday period may cost more than it is worth in terms of immediate payoffs. But, when demand is slack, the subcontractor may find itself faced with long periods of inactivity and no guaranteed income. Furthermore, the buyer firm needs to be sure that the subcontractor does not skimp on quality. If the subcontractor behaves opportunistically, and supplies substandard components, this can have an enormous impact on the overall quality of the machine, which will sometimes not emerge for some time, causing serious (and potentially expensive) delays, even if the faulty component can be returned for reworking. Finally, buyer firms may face a "hold up" problem from their subcontractors: if a final firm must rely on a subcontractor to produce a part that is vital to the production process, the subcontractor may use its position of power to extort rents from the final firm.

This problem is structurally similar to problems in the relationship between employers and employees, where employees may provide (nonobservable) effort if the conditions for gift exchange (generalized reciprocity) exist (Miller 2001). As in employment relations, the most obvious solution to this problem is a long-term relationship between the final producer and the subcontractor, in which the producer is prepared to pay a premium over the longer run to reward the subcontractor's commitment. In such a relationship, the subcontractor's short-term costs may be counterbalanced by longer-term profits deriving from the relationship, and the subcontractor furthermore has an incentive not to behave opportunistically insofar as the relationship will at some stage be opened for renegotiation. Such a relationship may permit some level of generalized reciprocity between the two parties, in which each party is prepared to accept short-term imbalances, given that the relationship is to its long-term advantage. Such reciprocity may be maintained between rational actors, "*as long as each party expects that the other party's continued participation is conditional on its own contribution.*"[19] However, creating the conditions for such reciprocity also involves a commitment problem on the part of the firm making the order. It has to be able to commit credibly to maintaining a long-term relationship in order to convince the subcontractor to cooperate in the short term where this is necessary.

Firms producing specialized machinery may choose either to produce specialized components internally or to put them out to subcontractors. If they make the first choice, they avoid the commitment problems involved in subcontracting,[20] but they must also themselves produce components that could be produced more efficiently and cheaply outside the firm. If they make the second choice, they may be able to secure components more cheaply, but they must face the problem of extracting credible commitments from their suppliers, as well as themselves making credible commitments to reward the

[19] Miller 1992, 204, italics in original. Gift exchange and reciprocity have typically been treated as nonrational; see Sahlins 1972; 1976; Polanyi 1992; and Trigilia 1998, but quite generalized forms of reciprocity can be maintained by rational actors; see the exemplary study of trust and reciprocity between Orma cattleherders and their agents in Ensminger (2003).

[20] For the purposes of this discussion, I wish to bracket the issue of commitment problems *within* the firm; see Miller 1992.

suppliers for their flexibility. Thus, creating the conditions for generalized reciprocity between subcontractors and final firms involves addressing problems of trust and cooperation. *Ceteris paribus*, it is difficult for final firms and subcontractors to trust each other enough to cooperate; each may have a short-term incentive to take advantage of the other.[21] However, if they can trust each other sufficiently, they can take advantage of some very substantial efficiencies indeed.

By examining whether business actors are able to solve the problem of trust and cooperation between firms in the production process, we may arrive at important conclusions about the factors underlying trust in industrial districts more generally. Where business actors are not able to solve the problem of trust, they will be more likely to prefer to produce essential components within the firm. Where business actors can trust each other enough to take the risks of cooperating through subcontracting, such cooperation will often be more rewarding than in-house production. Thus, we may expect to see quite different patterns of production and cooperation in situations where business actors are able to trust each other over this vital set of transactions than in situations where they cannot. A political culture explanation leads us to expect that firms will find it easier to cooperate in these areas where there is a high level of diffuse interpersonal trust. An institutional explanation, in contrast, leads us to expect that institutional variation will be the key explanatory factor. I examine the respective merits of these two accounts in the next section.

4.4 INTER-FIRM COOPERATION AMONG MACHINERY PRODUCERS IN BOLOGNA AND STUTTGART

There is clear evidence of differences in subcontracting patterns in the two case studies, which are furthermore linked to differences

[21] One might argue that the problems involved are sufficient to prevent subcontracting from taking place. This would be the likely perspective of Oliver Williamson's transaction-cost economics, which predicts that transactions with a high level of asset specificity (i.e., where the risks of opportunism are high) will take place within the firm rather than between firms (Williamson 1985). However, as Carlo Trigilia (1998, 392) argues, "in real life situations, ... even transactions with a high level of asset specificity may ... not be internalized [within the firm] if the institutional context limits opportunism and reinforces the bonds of trust."

in institutional settings. Italian interviewees made it clear that the judicial system was ineffective in providing enforcement; firms were reluctant to go to court even when they clearly had a good case:

> [legal action] is a waste of money [*spendere soldi per niente*]. Because in a legal action, one spends money on the lawyers, then it goes on, in practice, for two years, three years, four years, five years, ten years, then it never finishes.[22]

However, the packaging machinery district of Bologna, like other such districts, had a set of local rules – community institutions – that governed the behavior of final firms and subcontractors. As one Italian interviewee put it, a firm which does not cooperate

> comes to be excluded from the system. Slowly, one does not have credibility any more, which allows one to remain in the system. The world of packaging is very small – even if it is big it is very small. It is thus that it is, there are many personal relations that serve also sometimes to avoid situations which ... do not have to be solved through legal means. He who doesn't keep to this rule – these aren't written rules – well there you are! [*beh*!][23]

These rules turned on probity in personal relations between actors within the industrial district and involved the sanctioning of opportunistic behavior between subcontractors and final firms. As described by one interviewee:

> there is a saying in these parts that a handshake is worth more than a piece of paper with writing on it ... this is an approach which is particularly characteristic of someone from Emilia-Romagna. If someone steps out of line once, tries to be clever in some way ... there is a characteristic tendency which is particular to this area ... to cancel [the deal].[24]

Not only would opportunism lead to a cancellation of the immediate deal but there was likely to be a "freezing off" of relations extending into the future.[25]

[22] Firm Interview 17.
[23] Ibid., 8. I have translated the almost untranslatable Italian expression *beh*! as "well, there you are," its most neutral and general interpretation. Equally valid translations might be "so much for that" or "so much for him."
[24] Ibid., 16.
[25] Ibid.

These informal rules involved more than the simple punishment of opportunists by those whom they had cheated. Reputational mechanisms played an important part in the packaging district, so that those who had cheated would not only be punished by the injured party but also find it far more difficult to find future trading partners:

> This is a very small world. Reputation spreads in three days ... in three hours, not days. This is a small world, everybody knows everybody, and this is good in one sense, because if you have a good reputation, ... you can capitalise on it, because you are known and this is bad, because if you have a bad reputation and have done something wrong, everybody knows it, and they try to avoid you.[26]

This provided firms with a clear incentive to behave in a trustworthy fashion:

> from the point of view of the subcontractors a mistaken behavior, [with] a rather small yield, will become known very quickly in the productive system to the other customers, and thus it is easy to exclude this firm from the relations of production. In this case we have a transparency of the market which makes it much more difficult to maintain incorrect behavior.[27]

Here, it is clear that communication among actors played a very important role in ensuring that actors who behaved badly would quickly acquire a bad reputation. There is an unmistakable resemblance between the social "equilibrium" these informal rules led to and the forms of nonstate enforcement modeled by game theorists such as Avner Greif (1994), Paul Milgrom, Douglass North, and Barry Weingast (1990), and Randall Calvert (1995b). In these models too, communication within a community of actors may allow honest behavior in bilateral relationships, even if the number of actors is quite large. Like Avner Greif's Maghribi traders, Bolognese packaging machine manufacturers invest in gossip with each other, conveying information about their relationships with subcontractors. And like Greif's agents, subcontractors who are known to have cheated find it difficult to form new relationships with firms in the

[26] Ibid., 2. On the importance of reputation and the close-knit nature of the packaging industry in Bologna, see also Firm Interviews 1, 5, 7, 8, 13, 17.

[27] Interview with researcher at Nomisma (local economic research organization).

community and soon find themselves excluded from new business opportunities. Although, unlike Greif's models (which do not speak to uncertainty), these rules allowed for communication over situations that had not been anticipated by the involved actors, their underlying enforcement mechanisms were strikingly similar.[28]

The existence of these informal institutions, which mandated personal honesty in business relations, allowed producer firms to put work out to subcontractors. Furthermore, the final firms could commit to their subcontractors, and vice versa, without resort to formal contracts. Formally binding contracts were rare in the district; only a few of the very largest firms seem to have used them. As a rule, informal relationships between business actors were sufficient. In the words of a few of the firms themselves:

> The long-term things, I would say that there is nothing written. In practice, there is nothing like a contract. There is a tacit consensus, whereby we have work and make orders from these subcontractors.[29]
>
> ... with Italian subcontractors, or in the neighborhood, one has to work [*giocare*] on the personal and emotional plane. It is verbal, not all written down.[30]
>
> To have then the work of the artisan at the beginning of the year, one makes an accord. We ask for the best work. With the duty of doing the best work for us ... most times it is a verbal agreement.[31]
>
> There is no formalized contract, one works a lot on reciprocal knowledge, on reciprocal trust.[32]
>
> What is applied is not a specific contract, it is a personal relationship between an office which puts out work and a small artisanal organization which produces it.[33]

[28] Albeit not identical. As Avinash Dixit (2004) observes, real-life sanctioning is likely supported at least as much by a certain degree of economic irrationality (a willingness to engage in costly punishment without direct reward) as by the kinds of complex structures of incentives that game theorists model. On the beneficial consequences of irrational behavior (or even the possibility that one belongs to an irrational "type"); see further Elster 1989, Kreps 1990b.

[29] Firm Interview 4.

[30] Ibid., 16.

[31] Ibid., 13.

[32] Ibid., 10.

[33] Ibid., 9. See also 14 and 17 for further evidence on the lack of resort to formal contracts in subcontracting relations in the area. There is one interviewee (Firm Interview 8) who suggests that formal contracts are becoming more important than hitherto; see further Chapter 5.

Because of the reputational penalties that final firms would suffer if they reneged on their personal commitments, subcontractors could trust them enough to reciprocate in turn, refraining from behaving in an opportunistic fashion and providing flexibility in the short term in the knowledge that their commitment would be rewarded over the long run. The conditions for a quite generalized form of gift exchange, with extensive cooperation, were met, because informal institutions allowed actors to commit credibly to each other, and thus to trust in each others' credible commitments. More specifically, actors of one type (business actors in final firms) were able to trust actors of another type (subcontractors) over a wide variety of issues. To use the terminology set out in Chapter 2, actors of type *A* were able to trust actors of type *B* over a wide range of matters *C*. Each individual agent of a final firm knew that because he was representing a final firm (a member *a* of type *A*), and he was dealing with a subcontractor that formed part of a more general class of subcontractors (a member *b* of type *B*), and because informal institutions existed that led members of type *B* to behave in a trustworthy fashion, he could trust individuals of type *B* over an important set of issues, without the backing of a contractual agreement. Similarly, individuals of type *B* were able to trust individuals of type *A* to provide a long-term relationship. Thus informal institutions and their associated enforcement mechanisms provided the basis for trust and cooperation between final firms and subcontractors in the Bologna packaging machinery industry.[34] The result was a high degree of informal cooperation in the production process:

> I would say that the characteristic of this sector is this: an extreme interchange and collaboration between ... suppliers, and firms who produce the goods.[35]

These relationships of collaboration and exchange involved a very high degree of flexibility on the part of subcontractor firms. Several firms spoke of the willingness of subcontractors to work long hours for them at short notice:

34 Although see Chapter 5 for evidence that this set of informal institutions did not extend to all relationships between final firms and subcontractors; a small number of very large firms had little incentive to behave in a trustworthy fashion to subcontractors.

35 Firm Interview 9.

From this collaboration comes the availability on their part to work also 12 hours a day for a week, in order to finish a machine.[36]

... if we have a date 30 days away, or at the end of the month, I will telephone them and say, "Look, the piece has to come 15 days early, because something unexpected has turned up, and can you do it soon."[37]

One also has to have the possibility of working Saturday and Sunday if it is necessary. And we have obtained this on occasion, precisely thanks to these relations which we have with our suppliers in this network, which participate in our results.[38]

Furthermore, problems of "holdup" appeared exceedingly rare. Indeed, at least in some cases, final firms were prepared to order specialized parts from subcontractors with whom they had a relationship, without bargaining over price until after the piece had been produced. While this may sometimes have led to difficulties for final firms, these difficulties did not involve subcontractors opportunistically jacking up prices after the fact, but the uncertainties of trying to fix a final price with the customer for a particular machine without knowing the costs of all the inputs.

The relations of generalized reciprocity that were supported by these informal institutions furthermore facilitated a radical disintegration of the production process in the Bologna packaging machine industry. As described by a local economic research organization:

the production cycle in the mechanical sector is broken up into various phases of work, which leads to a high reliance on the flexible specialisation model. Because of this, we have systems in which the firms at the center are specialised in certain phases of production, solely assembly and planning, and all the phases of the production process are delegated to other firms, which are specialized.[39]

The production of specific components was usually carried out by a multitude of artisanal firms, each specializing in a particular phase of the production process, working on behalf of larger firms which sold the final product:

[36] Ibid., 3.
[37] Ibid., 1.
[38] Ibid., 16. See also 15, 17.
[39] Interview with Nomisma researcher.

little artisanal outfits that ... provide particular mechanisms [are] a Bolognese tradition. Firms do not produce one part of the goods that they may come to sell; they are made by subcontractors.[40]

The usual arrangement was thus one in which final producers designed the machines, sent the specifications for particular parts and mechanisms to smaller producers, and then assembled, sold, and maintained the final product. The process of assembly (*montaggio*) was more important than it sounded; indeed, it was perhaps the most strategically important part of the manufacturing cycle. Italian manufacturing firms usually did wish to retain some control over the process of ensuring that components manufactured by different firms worked well together. But firms sometimes put even this vital part of the manufacturing process out to others:

The situation is that we don't produce anything inside. We make everything outside – that is typical. We only do the planning and the assembly. And the testing. Then, not only [the above], we sometimes subcontract out the assembly, when we have a lot of requests. Thus, one also sometimes subcontracts out planning. Technical studies. Thus, the only thing that we don't want to subcontract is the sales.[41]

This disintegration of the production system captured certain economies that were not available in more conventional production relations:

[The system of production in Emilia-Romagna] has allowed one to have wide specialization with modest investments. Why? Because the artisan, who has acquired this machine for doing this stage of work, [does] 8 hours, 12 hours, 15 hours, 20 hours. One specializes in this sector, and amortizes the cost of the machine much more quickly ... In the German "system," between inverted commas, a big firm buys machines, and works to do this stage of work only for those pieces which it needs for its own production. Probably, this machine remains unused most of the time. Then, the amortization of costs is slower.[42]

As previously noted, this form of production was not merely a result of informal institutions mandating trust and cooperation; it was also

[40] Firm Interview 9.
[41] Ibid., 7.
[42] Ibid., 8. For a more general overview of this question, see Miller 1992.

a response to incentives provided by the Italian legal and labor relations environment, which made small-firm production a quite attractive option where it was possible. Nonetheless, even if the right kind of trust relations did not provide a complete explanation of this form of production, they were necessary to that explanation. Without high levels of trust and cooperation between firms, a radical disintegration of production would have been difficult or impossible to achieve, because of the high degrees of uncertainty in the production process and the consequent need for trust between final firms and subcontractors.

Thus, informal institutions in the Bologna packaging machinery district allowed actors credibly to commit to each other, and thus to trust each other enough to maintain relationships involving a high level of reciprocity and gift exchange. This in turn provided the necessary flexibility to allow firms to subcontract out highly sensitive and important parts of the production process, facilitating a quite extraordinary disintegration of production.

The machine tool industry of Stuttgart involved a rather different vertical organization of production. While formal institutions, in contrast to Italy, were effective and credible, there was no evidence of the kinds of extensive informal community institutions that characterized the Bologna machinery industry. Extensive subcontracting of the sort typical in Bologna was quite rare. One Italian interviewee, in an unprompted description, depicted the difference:

> The German system sees the firm, which is to say the unit of production, as an almost complete system, where one carries out all of the activities which form the product. In Italy, social organization is different, historically ... An industrial firm in Bologna – however, it is more or less like this in the rest of Italy too – plans this part for itself, which makes up part of a complete machine. It does the design, and then there is an entire series of external workers, to which it sends the design for getting the primary material. And then, it passes this on to another who does the first stage of work, and then onto another who does another stage of work; this is the procedure for returning a finished part, or perhaps [it is] already assembled into a small group which is mounted onto the principal machine.[43]

[43] Ibid. The interviewee had been asked a general question about differences between the Italian and German vocational training systems.

This dichotomy was overstated, but it did capture genuine differences between the two systems. While the decision of whether to subcontract or not did not solely depend on whether or not subcontractors could be trusted not to behave opportunistically, it was clear that it was substantially more difficult in Stuttgart than in Bologna to achieve the necessary kinds of generalized reciprocity. Of the relevant interviewees in Bologna, only three very large firms did not rely on extensive subcontracting, and two of those three still subcontracted out up to 70% of components at times of high demand (I discuss these firms further in Chapter 5). In Baden-Württemberg, just the opposite was true: only a very small proportion of firms engaged in extensive subcontracting (see below).[44] A large proportion of the firms interviewed in Baden-Württemberg preferred not to rely on outside subcontractors except for standardized inputs, either because of the time wasted in bargaining or because firms could not be relied on to deliver parts on time or to provide the necessary quality. One firm responded when asked whether it outsourced work:

> No, we have an in-house production of over 90%. This is clearly related to the fact that we face ever shorter times for our products, and therefore can't spend a lot of time negotiating over supplies – "Can you make me this, and what does it cost," and so on. That doesn't work.[45]

Another firm found that it could be "simply more flexible" if it relied on internal production rather than outside firms. It had found that it could not rely on subcontractors to deliver in a timely fashion: "it has sometimes happened that a supplier says 'I cannot supply this part this week any more.' And I need it urgently."[46] A third firm reported similar difficulties in getting subcontractors to deliver on time; it had also sometimes received unusable parts or ones requiring extensive remachining.[47] A fourth firm spoke of the problem of information

[44] One might reasonably object that this disparity may result from the sample size, which may lead to distorted or inaccurate findings. However, other, larger N statistical surveys support the broad picture of subcontracting practice suggested by my research results, even while they do not, by their nature, address the finer-grained questions of trust and cooperation discussed in this article. See, for example, Grotz and Braun 1997.

[45] Firm Interview 33.

[46] Ibid., 25.

[47] Ibid., 32.

transfer; it could not be sure that firms at a geographical remove would consistently implement technical changes to improve the product.[48] A fifth firm had outsourced work to a foreign firm but stopped because of quality concerns with the components that were being supplied. Now it produced all of its work in-house.[49]

Another group of firms went somewhat farther; they did outsource components, but only where such components were not critical to the firm's specialist strengths. One firm had recently begun a strategy of putting out as much work as it could, but only

> that part that is neither time-critical nor quality-critical, and that has no significant value-added for us. In this way, we want to concentrate on our core competences, and reduce our in-house production [*Fertigungstief*].[50]

Other firms had resorted to outsourcing during a period of crisis in the German machine industry between 1991 and 1993. This was sometimes simply a temporary response to short-term market difficulties:

> The trend previously was to work more with subcontractors. That was also naturally affected by the recession, in that one sought to keep as little in-house as possible, so as to reduce costs. I am rather positive to the opposite strategy, namely making a lot in-house. In other words, now that we are over the worst years I would rather take a risk and add an employee.[51]

However, in other cases the crisis had led to longer-term organizational changes; a very small number of firms had moved to a much more extensive dependence on subcontracting.[52] Clearly, this group of firms had managed, to some considerable extent, to create the conditions for generalized reciprocity. As in Italy, this involved the creation of long-term relationships with suppliers who provided critical components:

> For things of central importance, we have long-term relationships. When someone uses production parts that are of elementary significance for the

[48] Ibid., 35.
[49] Ibid., 20.
[50] Ibid., 21.
[51] Ibid., 35.
[52] Ibid., 18, 36.

machine, then one can't go on the principle of "Here today, gone tomorrow." It doesn't work. One then needs long-term relationships.[53]

We need those strong connections because we need to be as flexible as we are. We cannot work with subcontractors who are not willing to change designs shortly before delivery, whatever that means in detail. We cannot place an order four months before the delivery and not be able to change anything in between.[54]

Given the very small number of Baden-Württemberg-based firms that engaged in extensive subcontracting, it is difficult to make definitive assertions; however, the evidence suggests that there was a key difference between the kinds of long-term relationships found in Bologna and Stuttgart. The latter did not involve diffuse informal commitments but were instead based on formal contracts[55]:

That is why our subcontractors usually know the amount of work that they will have for us. We do not order individual components from them. We contract the amount of business for the next year or for the next two years or for the next three years, so they have already contracted all the work they will have for us for the next three years. The only thing which is not fixed is the date of the delivery for the individual components. So we have very long term contracts with companies, and that is why they are easily willing to help us when we need changes.[56]

These contracts did not spell out in detail the obligations of the two parties but instead provided a technology of commitment. The final firm could commit to a long-term relationship with the subcontractor, but the subcontractor had an incentive to provide flexibility insofar as the contract would at some stage in the future be opened up for renegotiation. Formal contracts thus served as a partial substitute for the informal institutions that governed subcontracting relations in Bologna. However, they carried a cost. Unlike informal obligations, they limited the final firm's ability to flexibly respond to changing market circumstances: it had formally agreed to take a number of components over a certain period. In Bologna, in contrast, final

[53] Ibid., 36.

[54] Ibid., 18.

[55] Here, my findings are supported by the more general literature on inter-firm cooperation in Germany; see Streeck 1992; Arrighetti et al. 1997.

[56] Firm Interview 18.

firms could deal more flexibly with changes in demand. As one Italian interviewee put it:

> There is always a turnover of work ... one never does contracts of one year, two years ... *Not ever with the difficulties [that there would be] for a firm of our size in making long term contracts with external suppliers.*[57]

This allowed Bolognese firms an "extreme flexibility [in the relationship], especially where the quantity of work is extremely reduced."[58] Thus, formal institutions, even when they provided the basis for reciprocal relations between final firms and subcontractors carried a cost, they tied the final firm to specific and formal obligations that carried into the future, even if conditions changed very substantially. Under some circumstances, formal institutions could help anchor a certain kind of trust between actors by allowing them to make credible commitments to each other and to build up trust of the kind that Hardin (2002) discusses, in which a long-term relationship provides an anchor for trustworthy behavior on the part of both actors. However, formal institutions did not serve as a source of trust so much as a means of removing one important cause of distrust (the risk for the subcontractor that the final firm would take advantage of its willingness to provide short term flexibility but refuse to provide a long-term relationship). The anchor for trust was not the institutions themselves, it was a one-to-one relationship (which was facilitated by the use of a formal institutional commitment mechanism, but not determined by it).

The absence of the kinds of extensive informal institutions found in Bologna meant that business actors in Stuttgart could not trust in each others' personal commitments as a means to secure reciprocity without further (and cumbersome) technologies of commitment. This had two effects. First, together with more general differences in labor market institutions, it meant that few firms in Stuttgart relied extensively on subcontracting, in contrast to Bologna, where it was the norm. Second, where such subcontracting did take place, it seemed to rest on formal rather than informal institutions, with important consequences for the forms of cooperation observed.

57 Ibid., 9, my italics.
58 Ibid.

4.5 CONCLUSIONS

The analysis of trust and cooperation in the two case studies has interesting implications for the comparative analysis of trust and cooperation. As I noted at the beginning of this chapter, industrial districts, even though Putnam (1993) cites them as evidence for the presence of civic values in Northern Italy, in fact pose a quite serious challenge to the political culture tradition of cross-national analysis. The existence of high-trust forms of production in a country with notoriously low levels of diffuse interpersonal trust suggests, at the very least, that the causal relationship between trust (as measured in mass opinion surveys) and economic behavior is rather more complicated than most political culture accounts would suggest. Claims (Putnam 1993; Inglehart 1997) that political culture or civic values can explain economic behavior (and consequently, respective levels of economic development) would seem to be misplaced, at least in these two cases. Observed variation in behavior is more or less the opposite of what a political culture account would predict.

In contrast, a careful examination of the causal processes involved suggests that the alternative explanation – differences in institutions – provides a much better account of the sources of observed variation.[59] This is *not* to say that institutions only affected cooperation through their effects on trust. The evidence strongly suggests that differences in labor market institutions too had important consequences for firms' decisions over whether to produce internally or to put out work to subcontractors. Italian firms likely had stronger incentives to put out work to subcontractors than their German equivalents, independent of the degree of trust between them.

However, a simple macroinstitutional account is not itself sufficient to explain observed patterns of cooperation; in particular, it cannot explain the high degree of variation in forms of inter-firm cooperation observed *within* Italy. As Burroni and Trigilia (2001) illustrate using geographic data, local production models based on small-firm cooperation play an important role in Italy, but they are

[59] The argument that inefficient legal orders are likely to be associated with a stronger role for informal mechanisms finds support in other contexts. For statistical evidence, see McMillan and Woodruff 2000. For a partial overview of this literature, see Dixit 2004.

not the only model. Other localities have different models of production, with far less reliance on subcontracting, and the key factor explaining this variation appears to be differences in local rules. Thus, even if one cannot (given problems of covariation), measure the respective contributions of industrial relations system incentives and other institutional factors to the observed outcomes, one can maintain that it is necessary to supplement a labor market institutions account with an account of the different relationships between formal and informal institutions in the two national systems. In particular, the existence of a commonly accepted set of formal, national institutions in Germany means that there is far less space for distinctive local institutional orders to emerge than in Italy, where such orders are commonplace. Some of these Italian local systems of informal rules, but by no means all, support quite extensive forms of trust and cooperation between business actors, that are difficult to achieve in a system that, like Germany's, relies heavily on formal, national rules.

This has particular implications for the case studies (although one should note that the Italian case can in no way be taken as a stand-in for the Italian economy more generally, given the wide variation among local economies in Italy). In the Italian case, the radical disintegration of the production process is the result not only of incentives provided by labor law but of the presence of informal institutions at the local level. These informal institutional rules made it rational for business actors to behave in a trustworthy fashion in subcontracting relations. Those who did not found themselves excluded from future opportunities to do business, not only with the party they had cheated but with other business actors in the community, who communicated among each other about their experiences in subcontracting relationships. Thus, institutions gave actors incentives to behave in a trustworthy manner and led to the dissemination of information that allowed actors to trust each other with regard to a specific – and highly important – set of issues.

In the German case study, where there were no extensive informal institutions of the sort found in Italy, firms typically were not able to trust each other enough to cooperate through the kinds of extensive subcontracting found in Italy. In the few cases where they did rely substantially on subcontracting, they apparently invoked formal contracts as a technology of commitment, which had important implications for the kind of cooperative relationship that could be sustained. In contrast to Italy, where informal institutions supported a highly flexible form of reciprocal gift exchange, which could shift according to changes in demand, German final firms, if they wished to make credible commitments to their subcontractors, had to do so through relatively inflexible contractual forms.

This variation involved an area of cooperation where we have good theoretical reason to believe that trust is important. If, as I have argued in Chapter 2, trust involves expectations over situations that cannot be precisely defined in advance (and thus cannot easily be handled by institution-induced confidence), then we may expect that appropriate informal institutions will provide a better means of anchoring trust than will appropriate formal institutions. A relatively diffuse set of informal institutions mandating honesty in personal relations, of the type that seems to prevail in Bologna, is far better suited to guaranteeing trust over future situations that are difficult to define than is a formal set of institutions underpinning the enforceability of legal contracts. The latter could, with difficulty, be adapted by business actors to foreclose certain kinds of opportunism but only by very substantially limiting actors' ability to adapt to circumstances that they had not anticipated.

In order to generate meaningful and nuanced causal arguments about causal relationships of this kind, it is not sufficient to observe gross statistical differences between cases. A detailed analysis of the relevant causal paths (Bates et al. 1998a; 1998b; Hall 2003) is the appropriate methodology for uncovering how complex strategic relations can generate observed outcomes (Hall 2003). By pursuing this strategy, one can explain how variation in patterns of trust and cooperation with regard to specific issues may occur against the direction predicted by those scholars who see diffuse interpersonal trust as an important explanatory variable. Diffuse interpersonal trust, far from being an immutable cultural orientation, may instead be viewed as the a kind of generalized *confidence* that is contingent on particular institutional arrangements (Jackman and Miller 2004) – effective formal institutions (Levi 1996; 1998). It may furthermore be distinguished from the kinds of trust and cooperation that are likely to occur in settings dominated by informal institutions, where one may expect deeper forms of trust and cooperation, albeit most likely among smaller groups of actors (Dixit 2004).

Thus, by studying institutions and cooperation, we may arrive at a better understanding of the difference between trust and institutional compliance. These two notions are typically confounded in both the theoretical and empirical literature, but they have quite different consequences for cooperation. So-called "high-trust" settings, such as Germany, involve a high degree of reliance on credible and effective formal institutions, which greatly reduce transaction costs by increasing individuals' confidence that others will behave honestly across a wide variety of transactions. However, institutions of this sort do not support trust directly (although they may have important indirect consequences for trust, by making certain kinds of opportunistic behavior less likely). Instead, they serve as a substitute for trust relations by allowing individuals to engage in more or less impersonal transactions.

For an important school of thought among economic historians, the economic success of advanced industrialized societies is a direct result of the move from highly personalized economic relations, through intermediary forms, to a world in which individuals may confidently engage in economic transactions with complete strangers (Greif 2005). The implicit or explicit emphasis of these scholars on the superiority of an economic sphere based on impersonal transactions to one based on personalized ones may quite possibly be justified. Impersonal institutions may massively increase the gains from trade in a world of heterogenous social groups and may be more socially equitable over the long term to boot (Greif 1994). However, informal institutions that support quasi-personal relationships, of the kind found in the Bologna mechanical engineering sector and in other industrial districts, can allow actors to achieve efficiencies and forms of cooperation that are difficult or impossible to achieve in a world of formal institutions and impersonal economic relationships. If the arguments canvassed in Chapter 2 and evidence from the two cases examined in this chapter have wider application, then appropriate informal institutions are better suited than formal institutions for anchoring trust in hard-to-predict situations. Precisely because appropriate formal institutions are capable of supporting confidence, they are likely to do a poor job at underpinning trust.[60] To use H. L. A. Hart's terminology, while they have very well-defined cores, they are likely, by the same token, to have quite narrow penumbrae. Appropriate informal institutions, in contrast, are likely to be rather less effective at guaranteeing confidence; they will be more diffuse and hazily defined than formal institutions. Equally, they are more easily applied in situations that were not precisely anticipated. Thus, in situations where trust rather than confidence is called for, they will provide a better underpinning for the kinds of diffuse reciprocity that David Kreps (1990a) and Gary Miller (1992) suggest are sometimes vital to economic success.

This suggests that there are some important trade-offs between formal and informal institutions. To the extent that the spread of formal institutions has indeed vastly expanded opportunities for economic exchange, it has surely been essential to the economic success of advanced industrial democracies. Nonetheless, Italian industrial districts such as the Bolognese packaging cluster have apparently succeeded in creating economic efficiencies that appear not to be available to producers in an environment with effective formal institutions. Thus, informal institutions, at least under certain circumstances, may have considerable economic benefits.

The relationship between informal institutions and increased trading opportunities is an important one for debates over industrial districts. Current arguments over the future of this form of economic organization (Berger and Locke 2001, Whitford 2001a) focus on whether or not districts are likely to survive the impact of globalization. If, as I have argued, institutions are key to explaining cooperation among firms in industrial districts, then accounts of change in patterns of cooperation within industrial districts need to focus on institutional change as a key intervening variable. I turn to this question in the next chapter.

[60] This judgment is relative; formal institutions in jurisdictions (such as Germany or Japan) where the principle of good faith is an important guide to judicial decision making in commercial disputes are likely to have at least a little bit more flexibility. See further Arrighetti et al. 1997.

5

Accounting for Change in Informal Institutions

The Dynamics of Industrial Districts

5.1 INTRODUCTION

Chapters 3 and 4 provided a comparative account of the relationship between institutions and trust. This chapter turns to a different task – that of explaining change in patterns of trust and cooperation over time. In particular, it seeks to explain how *changes* in informal institutions may lead to changes in patterns of trust. As discussed in Chapter 4, informal institutions of the right kind can support thick forms of trust among actors, while effective formal institutions are more likely to produce institutional compliance than trust as such. This suggests that one very important mechanism driving changes in trust and cooperation among economic actors is likely to be a change in the informal institutions that support them.

These specific arguments about changes in informal institutions offer insights into more general debates about how and whether national and local institutions are being affected by exogenous shifts in the international economy. As local and national economies become increasingly interpenetrated with each other, actors' opportunity structures will change: some of them will begin to take advantage of new external options that were previously unavailable to them.

Although the consequences of these changes for existing institutions have received considerable attention in the literature, they are still the subject of controversy. Many authors have claimed that cooperation-intensive forms of production are being corroded by the logic of an

Anglo-American style free market economy (Ohmae 1990; Greider 1997; Friedman 2000). However, not only does much of this work lack either rigor or properly developed arguments about the relevant causal mechanisms but it also fails almost completely to take proper account of how institutions may intermediate change (Campbell 2004). While there is an important body of work that seeks to counter overly broad claims about global market change by showing how different institutional structures may adapt to new challenges rather than simply being swept away (e.g., Campbell 2004; Hall and Thelen 2007), this body of work in its turn pays relatively little attention to how *informal* institutions – the kinds of institutions most important to industrial districts – are likely to change.

Thus, even though informal institutions play an important role in cooperation intensive economies, there is little theoretically grounded work that seeks to address how they may alter as a result of broader changes in global markets. Academics have focused on change in formal institutions (which is easier to demonstrate and to quantify). This begs some important questions. Do informal institutions change in the same ways as formal ones when background conditions shift? Are the mechanisms of change similar or different? On the basis of the existing literature, it is impossible to say.

Italian industrial districts – and the trust relations between final firms and subcontractors within these districts discussed in the previous chapter – pose these questions in an especially pointed way. Recent debates over industrial districts have focused exactly on the question of whether or not the informal relations that have characterized subcontracting relations in these districts can survive globalization. Some scholars argue that the forms of trust and cooperation found in industrial districts will wither away as a result of changes in global markets, while others suggest that they are likely to continue to survive and perhaps to prosper.

On one side of this debate, prominent authors in the globalization literature (e.g., Amin 1993; Harrison 1994) have predicted that economic systems (such as industrial districts) that are based on localized cooperation will be corroded as key firms come to prefer global economic relationships over local ones and as large firms from both inside and outside the districts in question reshape local economies better to their liking. This leads authors such as the late Bennett Harrison

(1994) to suggest that the informal patterns of cooperation that characterize subcontracting relationships in industrial districts are likely to wither away as large group firms come to predominate and to reorganize inter-firm relations on the basis of hierarchy. Harrison's widely read monograph on corporate power and change uses the Bologna machine producing district described in this book as its key example of how increased hierarchy, centralized decisions over subcontracting, and formal contracting relations go hand in hand to weaken traditional forms of trust and cooperation. He describes how large firms have bought out smaller final firms in this district and assembled them into groups (in Italian, *gruppi)*, which seek to take advantage of economies of scale and organization through common sourcing of components and the imposition of formal contracts on subcontractors. In his words the Bolognese machinery industry is an example of how

> Decisions about where to produce, which technologies to employ, and which subcontractors are qualified; the freedom to deal directly with customers; and the locus of industrial relations have all been disrupted by the penetration of the districts [by large outside firms]. (Harrison 1994, 87)

This leads Harrison to doubt whether "cooperative competition, trust, and the other underlying principles that have been identified as the glue holding the distinctive Italian geographically based production networks together" can survive (88).

Harrison's arguments receive support from many Italian scholars. For example, Maura Franci and Vittorio Rieser (1991) argue that industrial districts are moving from a model where economic relations depend on customs of informal reciprocity to one characterized by more formalized relations, where some firms become more dependent on others. So-called *aziendalisti* – scholars who argue against a previous generation of research that emphasized the collective aspects of industrial districts, and who base their explanations on the strategies of individual firms (Whitford 2001a) – also tend to be highly skeptical about whether districts can survive unless large firms restructure them.

These theoretical claims go hand in hand with a host of empirical studies which show that districts are becoming more hierarchical in response to global pressures on demand markets. Thus, for example, Francesco Brioschi, Mario Sole Brioschi, and Giulio Cainelli (2002) use the experience of Bologna and other industrial districts to argue that the "widespread formation of groups in industrial districts" is leading to "more hierarchical relations between firms," and the replacement of flexible relations based on informal cooperation and trust with more stable and less egalitarian relationships. As described by another Italian academic:

> The Marshallian district is a district where power relations are such that no-one is really the big power – even the *impannatore* in Prato is a strategic organiser in the system which is not so powerful, not hierarchically different from the other companies, part of the community. And this is what is changing in the districts. In some districts it has changed dramatically, very clearly, [such as] Sassuolo and Biella.[1]

These arguments typically go together with a set of claims about the irrelevance or obsolescence of informal institutions. *Aziendalisti* typically either discount informal institutions in favor of firm-level factors or argue that these institutions are likely to wither away in favor of more formal kinds of inter-firm hierarchy (Ferucci and Varaldo 1993; Cainelli 2002).

On the other side of this debate, scholars such as Suzanne Berger and Richard M. Locke (2001) provide a more positive account of how the traditional virtues of industrial districts apply in a changed global setting. Like Harrison and his colleagues, they see industrial districts as vital test cases for the ability of cooperative production systems to survive global change, arguing that

> if it turns out that even in industries like garments and ceramic tiles, that Italian small and medium scale district-based enterprises can prosper in global competition, then we need to revise expectations about the vulnerability of territorially-embedded economic arrangements to the mobility of capital, goods, and services across borders. (Berger and Locke 2001, 93)

Unlike Harrison, Berger and Locke argue that districts *have* weathered these changes successfully. They find "scant evidence" that industrial districts have seen any general increase in hierarchy in inter-firm relationships, although they do note a trend toward greater vertical integration. Nor do they find that there is significant

[1] Interview with Nicola Bellini.

evidence of penetration by outside firms. This leads them to conclude that embedded networks of cooperation and "local interdependencies and horizontal ties [continue] to be the norm" (96). More generally, they conclude that "what is evident in the districts today is a pattern of adaptation that builds on the 'old economy' and does not displace it" (101).

Other scholars adopt a somewhat more circumspect position, claiming that industrial districts are changing more than Berger and Locke suggest, but that they are not in a general state of decline. For example, Burroni and Trigilia (2001) suggest that there *is* clear evidence that industrial districts in Italy are becoming more diverse in their modes of internal organization, so that the canonical "network of firms" model in which no single firm or small group of firms predominate is being supplemented (and perhaps in some instances supplanted) by "networked firm" models in which a small group of firms and their suppliers play a more important role. This has been reinforced in Italian regions such as Emilia-Romagna (Farrell and Holten 2004) by regional economic agencies, which are moving away from a generalized model of small firm support to focusing targeted assistance on a small group of "leader firms." However, contrary to the claims of Harrison et al., these scholars argue that changes in ownership structures, and the consequent increase in market hierarchy, do *not* seem to be undermining diffuse cooperative norms. A greater degree of hierarchy in relations among some firms seems to exist side by side with continued reliance on informal cooperation in subcontracting relations.

In order to adjudicate better between these disagreements over how globalization affects trust and cooperation in industrial districts, we need a better account of the intermediating role, if any, of informal institutions. Scholars such as Harrison and the *aziendalisti* provide an account of change in which local informal rules are at best epiphenomenal. For them, the important causal pathways run through market decisions taken by firms (and in particular by large firms) without reference to local rules.

In their account, businesses react to the opportunities and constraints of globalization in a more or less untrammeled way: when the international economic environment provides them with new ways of making profits, they naturally take advantage of them. Most particularly, globalization provides large firms with the opportunity to take advantage of economies of scale by buying out their smaller competitors and hence imposing a new, formalized, and rationalized set of subcontracting relations on the district (Harrison 1994). It also allows these firms to forge new relationships with a variety of both local and nonlocal firms. As a result, we may expect localized forms of trust and cooperation to diminish and wither away in favor of more formalized relationships between large-firm dominated *gruppi* and a smaller group of subcontractors. In this account, local informal institutions are irrelevant to changes in firm behavior. Instead, the key intermediating factor between global market shifts and changes in subcontracting practices is *change in ownership structures* among final firms, under which large final firms buy out smaller final firms, integrate them into tightly knit groups, and reshape subcontracting relations in the district as a whole.

More traditional accounts of industrial districts disagree, arguing that actors in industrial districts *are* constrained by local informal rules and that these rules are more than mere epiphenomena. As previously discussed, there is broad agreement in the traditional literature that trust and cooperation – insofar as they applied and continue to apply in subcontracting relations – are based on community rules of the sort discussed in the previous chapter. However, even while this literature continues to maintain that informal institutions are important, it has not to date laid out a grounded theoretical account of how these institutions are likely to respond to globalization and similar external shifts in markets. Thus, if we want institutionalist accounts to explain how globalization affects trust and cooperation among firms in industrial districts, we need a better account of how informal institutions are, or are not, likely to respond to global economic pressures. In short, institutionalists need to incorporate the *specific mechanisms of institutional change* into their causal accounts if they wish to provide a plausible counterexplanation of how and whether firms' cooperative strategies will change or remain stable in response to changes in global market conditions, and they need to test their account against alternative explanations.

This not only has relevance for specific arguments about industrial districts but also for a much wider debate about institutional change and the global economy. In this sense, industrial districts serve, just as Berger and Locke describe them, as an important test case for a much broader set of arguments. Precisely because subcontracting relations in industrial districts (at least in their Italian incarnation) rely so heavily on informal rules, they allow us to single out more precisely the causal pathways through which global economic changes can affect informal institutions, and to begin to determine which of these possible chains of causation appear to accord best with the available empirical evidence.

Thus, by examining change in industrial districts we can accomplish two tasks. First, as I discuss above, we can better come to understand the forces affecting continuity and change within these districts, a topic that is of considerable interest in and of itself. Second, we can begin to articulate a much-needed more general theory about the relationship between economic change and informal institutions. There is already a quite well-developed literature on the relationship between globalization, formal institutions, and economic behavior. However, relatively little attention has been paid to how informal institutions mediate the relationship between the global economy and the expectations and behavior of specifically situated economic actors. A specific account of the mechanisms of informal institutional change would supplement (and to some extent redress imbalances in) our accounts of how the formal institutional elements of the political economy mediate global pressures. We cannot understand the circumstances under which informal institutions do or do not respond to changes in their external environment, nor, more broadly, the behavior of economic actors subject to these institutions, without a proper understanding of these mechanisms. However, scholars neither of international political economy nor of comparative political economy have succeeded in developing an account of the relationship between international market changes and domestic informal institutions.

Here, I lay out a theory of informal institutional change and show that it can help us understand how informal institutions undergirding trust can withstand changes in external environmental conditions. I begin by providing an overview of the relevant literatures in both international and comparative political economy. I build on insights from these literatures to create a theory of informal institutional change. I then apply this theory to explain how the informal institutions supporting cooperation in the Bologna packaging machinery district and in other Italian industrial districts appear to have survived broad changes in power relations and forms of economic organization. I conclude by examining the relationship between informal institutional change and the broader kinds of institutional change discussed in the literature.

5.2 INTERNATIONAL MARKETS AND DOMESTIC INSTITUTIONS

How do shifts in the international economy affect informal institutions, and thus cooperation and trust? The existing literature provides little guidance – as Helmke and Levitsky (2004) note, there is little current empirical work in political science on how informal institutions emerge and/or change over time. Political economists' arguments about informal institutional change are especially scanty.

International political economists usually fail to consider informal institutions at the domestic level at all. International political economy (IPE) scholars interested in the domestic economy have instead focused on explaining how market shifts affect the state's preferences over regulating economic activity. Many of the causal arguments put forward by scholars of international political economy hardly touch on institutions. Instead, their theories of state and social group preference formation invoke arguments based on Stolper-Samuelson (Rogowski 1989) or Ricardo-Viner (Frieden 1991) theories about how preferences emerge among interest groups and thus come to structure cleavages in domestic politics. Arguments in this vein see government policy as a simple response to interest-group politics and pay scant attention, if any, to institutions (Garrett and Lange 1995). Although more recent work has explicitly built institutions in (Garrett and Lange 1995; Milner 1997), it has concentrated exclusively on the formal institutions of the state (Frieden and Martin 2002).

Comparative political economy provides a much more promising source of potential insights. Scholars in comparative political economy, and most particularly in the "varieties of capitalism" literature have a long-standing interest in how institutions – including informal ones – may support cooperation among economic actors. The most influential recent book in this genre, Peter Hall and David Soskice's (2001) *Varieties of Capitalism* (see also Soskice 1999) argues that one can distinguish between two internally coherent forms of national economic organization: the "coordinated market economy," in which firms organize themselves through a variety of nonmarket mechanisms, and the "liberal market economy," in which firms organize themselves through market relations and (internal-firm) hierarchy. In Hall and Soskice's framework, "culture" and "informal institutions" embody "shared understandings," resulting from "repeated historical experience [which] builds up a set of common expectations that allows the actors to coordinate effectively with each other" (2001, 13) in coordinated market economies.

These arguments have much common ground with those laid out in this book. However, they also lead Hall and Soskice to treat informal institutions as an ancillary means of smoothing out potential difficulties that formal institutions cannot themselves resolve and conducting actors toward a mutually beneficial equilibrium. While informal institutions are not *quite* construed in functionalist terms, they are depicted as being primarily important insofar as they help support broader forms of coordination within the economy in question.[2]

More recent scholarship, which seeks to recast the varieties of capitalism literature in more dynamic terms by wedding it to a theory of institutional change, shares this tendency. In work under progress, Peter Hall and Kathleen Thelen (2007) go some considerable distance toward answering critics (Crouch and Farrell 2004) who have complained that the approach is overly static and/or implicitly functionalist in its reliance on arguments about institutional complementarities. Hall and Thelen argue that the boundaries of institutions are, in fact, continually contested by self-interested actors, who are perpetually engaged in an "ongoing reassessment of both their own scope for action and the intentions of those with whom they are interacting," engaging in conflict "designed to test the limits of cooperative arrangements" or mobilizing others into line with institutional arrangements. Here again, common knowledge, built up over repeated interactions between actors is important; it makes it less likely that cooperation will break down. Hall and Thelen identify three key mechanisms that may lead to institutional change – reform, defection, and reinterpretation.

[2] For a valuable corrective, and the seeds of an alternative approach, see Hall 2005.

Even though this work substantially advances our understanding of institutions, it does not, as currently constituted, provide a very good account of how informal institutions (as distinct from formal institutions) are likely to change. Hall and Thelen focus on change in formally negotiated institutions; the empirical examples that they draw on are wage bargaining, works councils, financial regulatory reforms, labor market reform, pension policy, corporate governance, and collective bargaining over work conditions.[3] While two of the three mechanisms that Hall and Thelen discuss (defection and reinterpretation) are likely to apply to informal institutions as well as formal ones, they are likely to operate quite differently with regard to informal rules (which are the product of decentralized bargaining) than to the formal rules that Hall and Thelen focus on (which are the product of more or less formal processes of negotiation). Finally, their account of institutional change has some ambiguities: it is not entirely clear which mechanism is likely to apply when, nor is it clear how the broader features of institutional change that they are interested in (perpetual contestation, power relations, and the like) intersect with these mechanisms. While Hall and Thelen's framework offers some highly valuable insights, it needs to be reconstructed or (more likely) supplemented if it is to provide insight into informal as well as formal institutional change.

The literature on informal institutions offers two plausible bases for such a supplementary account: one that emphasizes how ideational factors lead to joint belief shifts and hence informal institutional change (Culpepper 2005), and one that emphasizes how power relations may produce such shifts (Knight 1992; 1995). The differences between Culpepper's and Knight's respective approaches reflect what Kathleen Thelen (1999) identifies as an emerging divide between materialist and normative approaches to institutional analysis. While I build on Knight's materialist account here, I return to the relationship between these two accounts in the concluding section of this chapter.

As Knight (1992) argues, not only do informal institutions instantiate beliefs about how different classes of actors are likely to behave but they are also likely to reflect the relative bargaining power of actors

[3] Some of these examples do involve informal reinterpretation of formal rules (Farrell and Héritier 2003).

belonging to these classes. Bargaining power, in this context, depends on the options that actors have should they fail to coordinate – those actors that have many attractive options will ipso facto be less vulnerable to failure than others without such options and hence better able to issue credible threats. Stronger actors will thus be able to force other, weaker actors to coordinate on solutions that are distributionally advantageous to the stronger actors, even if they are relatively disadvantageous to the weaker ones.

As similar (but by no means necessarily identical) situations arise again and again between particular classes of actors, members of these classes will come to recognize the likely outcomes, and the actions that are "appropriate" (in the most morally vacant sense of that word that is possible) for all actors belonging to the relevant classes in these situations. This is how informal institutions emerge – sets of rules that are never formally negotiated through some collective process, and that are not enforced by an external specialized actor such as the state or court system, but that may nonetheless be pervasive and have quite important consequences for actors' behavior.

It is important to note that this process of informal institutional emergence is neither mechanical nor automatic. Indeed, in practice it will involve precisely the kinds of ongoing reassessments of interest and conflicts over how to interpret institutions that Hall and Thelen describe. Conflicts over informal institutions, however, are likely to play out very differently than conflicts over formal ones. They will take place neither through formal negotiations nor through competitive efforts to influence the interpretations of existing institutions offered by specialized adjudicators such as judges. Instead, they are likely to be spurred by specific actions taken by initiating actors, which deviate from the common understandings instantiated in an informal institution, "pushing the boundaries" of this institution in one sense or another, or even deviating from it completely.

This means that the key challenge faced by accounts of informal institutional change is to explain (1) the circumstances under which these actions will or will not arise and (2) the circumstances under which they will or will not be translated into the *generalized* changes in expectations over mutual behavior among different classes of actors that we associate with institutional change. In contrast to formal institutions (whether they be state regulations or judicial rules), informal institutions offer no point of centralized decision making that actors may seek to capture in order to push through change. Instead, whatever changes occur will happen through a decentralized process, in which multitudes of actors belonging to particular classes change the ways in which they interact in repeated situations of a broadly similar type.

This suggests that informal institutional change will be likely to happen if three conditions are satisfied. The first condition is that the action taken by an initiating actor (an action which deviates to a greater or lesser degree from an existing informal institution) leads to higher payoffs for the actor who has undertaken it than would be achieved under the status quo. Obviously, if the action does not lead to a better individual outcome, we may expect that the initiating actor (and actors in her broader class of actors who might learn from her) will not repeat it in future, and that it hence has little chance of becoming institutionalized.[4] The second is that this action can be sustained and repeated over time by the initiator, in the face of potential opposition from any actors who are disadvantaged by it relative to the existing dispensation. The third is that this action can be taken up by most or all other actors belonging to the same general class as the initiator, so that everyone in the relevant community comes to assume that actors belonging to that class will resort to that action in situations of this kind in future, and recalibrate their own strategies accordingly.

Under these circumstances, bargaining power, and the respective sets of options available to different classes of actors, will surely shape informal institutions, as Knight and others argue. Informal institutions will respectively change or collapse when classes of actors with sufficient bargaining power are able and willing to make credible

[4] It is plausible that under some circumstances, an action or accumulation of actions deviating from the existence of the institution may lead to the collapse of an institution, even if they are not to the initiator's advantage. For example, if the consequences of the action inadvertently reveal to other actors that *they* would be better off in distributional terms without a given institution, these other actors may seek to systematically deviate from that institution in the future. However, fully exploring the implications of this would require a much more complex account of institutional change than I am able to provide here (although I hope to pursue it in future in joint work with Cosma Shalizi); I therefore bracket this possibility for the purposes of current discussion.

threats so as to reshape an institution (e.g., to threaten that if they do not get their way in particular bargaining negotiations, they will explore other attractive options), or have outside options that are so attractive that they decide to deviate so completely from an existing institution that it falls into utter disrepair.

As noted above, actors' bargaining strength depends on the extent to which there are *attractive* options available to them in the case of breakdown. "Attractive," in this context, means options that are roughly equivalent to or better than continued cooperation under the existing set of understandings.[5] In other words, unless the putative benefits of change are greater than, or equal to, the benefits of continuing to work under the existing dispensation, the threat of change will not be credible. Nor, obviously, will actors seek to deviate from the behavior set out in existing institutions unless they believe that it is to their individual advantage to do so.

This means that we may expect to see exogenous shifts in circumstances (such as those resulting from globalization) to result in informal institutional change *only* (1) when they provide opportunities of action to individual actors that are equivalent to, or better than, those they are likely to enjoy under the existing dispensation in terms of their payoffs, (2) when these actions can be repeated and sustained over time, and (3) when the other actors belonging to the general class of actors in question can take up and sustain these actions too, so that these actions become the "expected" response of the general class of actors in a given situation.

However, we may also expect that individual actors will not have complete knowledge over how exogenous shifts in circumstances (such as, in the case under discussion, those attendant on globalization) may have changed both their bargaining power and the outcomes that they can reach. Thus, any moves they make to change the status quo will be

[5] Note that "attractive" is a relative term rather than an absolute one. We can thus imagine two ways in which exogenous change may make actors want to take new actions. First, exogenous shifts may lead to previously unconsidered actions becoming more attractive while the *status quo* remains the same. Second, exogenous shifts may lead to the status quo becoming considerably less attractive than it had been, leading actors to want to consider new actions. While these will be associated with different empirical situations (the first involves exogenous change revealing new opportunities, the second involves exogenous change threatening the old ways of doing things), they both can be explained by a common theory.

made under uncertainty (the initiating actor or actors may hope that these moves will improve their payoffs, but they cannot know it).

This means that we may expect that many actions that deviate from existing informal institutions will be "mistakes" that will not be repeated, either because they turn out not to increase payoffs as anticipated or because they encounter successful opposition from other actors. "Mistakes" of this kind will not lead to informal institutional change. Furthermore, some actions that are not "mistakes" may depend on idiosyncratic features of the initiating actor, so that they prove difficult or impossible for most or all other actors in the general class to which the initiating actor belongs to emulate. These actions too will fail to result in broader institutional change, because they will not change the broader expectations held within the community about the relevant class of actors. In other words, unless the action in question is adopted as a general response among the class of actors in question in relevant situations, it will not change expectations regarding that class, and hence it will not change relevant informal institutions.

How might this theory be applied to questions of informal institutional change in industrial districts? At this very broad level, the theory does not provide specific predictions. In order to advance hypotheses, one needs to have a better specified theory of the *particular* sources of bargaining power that different classes of actor (in this case, final firms and subcontractors) have, and how these actors' relative bargaining power are affected by exogenous factors such as globalization.

The beginnings of just such a theory are provided by my arguments in Chapter 4. Building on these claims, one may posit that the relative bargaining power of final firms and subcontractors is in large part conditioned on their degree of mutual dependence. As Chapter 4 discusses, most final firms and subcontractors in the packaging machinery district of Bologna (and of other industrial districts) are highly mutually dependent on each other – final firms are structured so as to rely extensively on subcontractors' willingness to provide informal cooperation, while subcontractors rely on final firms' guarantees of continued work. The informal institutions that shape their mutual interactions provide them with considerable rewards (the benefits of flexibility) while limiting risks of mutual opportunism.

This in turn limits the degree to which shifts in external circumstances will necessarily increase the bargaining strength of either final firms or subcontractors. Globalization, and the new opportunities and challenges attendant on it, will only result in changes to informal institutions when it provides options to one or more classes of actors that are as attractive, or more attractive, than the options that they have under the existing set of informal institutions. Given the very considerable efficiencies that most final firms and subcontractors enjoy under the existing dispensation, and the real risk that these efficiencies may unravel if one or the other set of actors deviates from the behavior that supports them, the relevant informal institutions will enjoy a reasonably high degree of stability – the opportunities of globalization will have to be quite enticing (or the threats of globalization very worrying) for actors in one or the other class to want to shift from the one equilibrium to another, and to be able to make credible threats that they will do so.

This means that these arguments provides us with a set of predictions that differs from those offered by Franchi and Rieser (1991), Harrison (1994), Brioschi, Brioschi, and Cainelli (2002), and others who emphasize the causal importance of ownership structures. These authors suggest that the tendency of large firms to buy out their smaller competitors and create "groups" is the key vector of change that is weakening traditional structures of cooperation. They thus claim that the creation of groups is likely to lead to common organizational practices among final firms and a much greater reliance on contracts and other formal mechanisms. Under their argument, we may expect groups of firms to create common structures for handling subcontracting relations through formal contracts, allowing them to take advantage of the economies of scale that are possible if firms within the group band together to make orders (Harrison 1994). This will eventually lead in turn to a more generalized collapse of informal patterns of trust and cooperation between final firms and subcontractors within the district.

These arguments would lead one to expect that the behavior of *both* large leader firms *and* the smaller final firms that they have bought and integrated into groups will reflect these changes. Both groups of actors will use formalized contracts with their subcontractors rather than traditional kinds of informal trust-based cooperation. If changes in ownership structures (i.e., the creation of *gruppi* and assimilation of smaller final firms within them) is the key factor leading to more formalized hierarchy and the weakening of traditional forms of cooperation in subcontracting relations, then we would expect that both the leader firms and those firms that they control through their *gruppi* will rely on formalized relations rather than informal ones in order to organize subcontracting.

These arguments have more ambiguous predictions for those final firms that are not integrated into groups. While Harrison and others suggest that large-firm strategies will likely lead to a general collapse of trust and cooperation among all firms in industrial districts, they are vague as to the mechanisms through which this is supposed to happen (Whitford 2001b). Plausible mechanisms might include extinction (independent final firms employing traditional forms of subcontracting might find themselves squeezed out of existence by larger ones) and emulation (independent final firms might seek to switch to formal contractual relations with subcontractors because they see these relations as one of their large firm competitors' sources of advantage). Depending on which mechanism is invoked, and how precisely it works, a considerable period of time might elapse before it had observable consequences.

A bargaining-power account of informal institutional change, emphasizing the factors that I have discussed above, would lead to quite different predictions. It would start from the position that actors are unlikely to take action that might challenge existing institutions in the first place and to repeat it, unless these actions promise benefits that are greater than or equal to the status quo and they have the bargaining strength to sustain it. It would also suggest that these actions are unlikely to precipitate institutional change unless they come to be generally expected behavior across the broad class of actors in question.

In the particular case of industrial districts, the specific willingness of final firms to take action that might challenge existing institutions governing subcontracting relations will in large part depend on the degree of mutual dependence between them and their subcontractors. Firms that are highly dependent on their subcontractors will both be less willing to undertake actions that might deviate from existing informal institutions (they are less likely to find alternative ways of doing business attractive) and less able to sustain these actions over time should they prove controversial (they will have less bargaining strength vis-à-vis subcontractors).

There is significant variation in the degree to which final firms are dependent on their subcontractors. A small number of very large final firms (precisely those firms that have been most active in buying up others) have substantial internal production resources, including most prominently a high degree of technical prowess and internal production capacity. Thus, unlike other firms in Bologna, they forgo some of the advantages of flexibility in favor of more traditional production structures. With the exception of a few very highly specialized processes, these firms are far more technically advanced than their subcontractors. While very large firms employ subcontractors as a means of responding more flexibly to unexpected fluctuations in demand, they typically turn to them only when their internal production facilities are working at full capacity.[6]

This means that even though subcontractors sometimes play a significant role for the few very large final firms in the district, they are not crucial to their production model. Subcontractors therefore have little bargaining leverage with very large final firms – they have few means to threaten holdup (Williamson 1985) and every reason to fear that if they do so in a specific instance, they will find themselves frozen out of relations in the future. Large firms can thus explore new actions deviating from existing informal rules and repeat these actions over time more easily – they neither benefit greatly from these rules nor depend as much on friendly relations with their subcontractors as other firms in the district.

A bargaining approach would suggest that the situation is very different for the vast majority of final firms in the district, regardless of their ownership structures. These firms are much smaller. They typically design and assemble machines but do not themselves produce parts, and they have neither the relevant technical knowledge nor the facilities to do so. This means that these smaller final firms are highly dependent on subcontractors – contracting out production is at the heart of their business model. They are thus likely to have neither the desire to challenge existing informal institutions, which

[6] Firm Interview 15.

structure the relations on which they depend for their livelihood, nor the ability to sustain challenges over time in the face of opposition from their subcontractors. Subcontractors who feel that they are not receiving the benefits of their implicit bargain (long-term guarantees of continued work) will credibly be able to threaten to withhold flexibility and additional effort, with serious adverse consequences for the final firm.

Thus, a bargaining account would predict, just as industrial district skeptics would, that large leader firms are relatively more likely to take actions that are at variance with existing informal rules and to sustain these actions over time. Since these large firms are less embedded in relations of mutual dependency with their subcontractors than other firms in the district, they will accordingly be more likely to deviate from informal institutions so as to respond to possible opportunities (or threats) in global markets. Simply put, these firms have much less to lose from experimentation (because the relative benefits that they enjoy from the existing dispensation are lower, they do not need outside options to be as attractive before they can use them as bargaining leverage).

However, a bargaining account would argue, in contrast to ownership structure accounts, that smaller final firms that have been bought by larger firms should behave similarly to smaller final firms that have remained independent. Smaller final firms that have become members of *gruppi* will still confront a situation of close mutual interdependence with their subcontractors, unless they are reorganized radically in ways that would undermine the sources of their competitive advantage. Except where outside opportunities or threats point toward very substantial gains or losses, the actors controlling them will neither want to take actions that challenge existing informal institutions nor (if they do so by mistake) be able credibly to make threats that would allow them to sustain these actions over time, since they will have insufficient bargaining strength vis-à-vis subcontractors. Because they face the same basic situation of mutual interdependence as independent final firms, they will behave in similar ways.

In short, ownership-structure accounts suggest that both large leader firms *and* the small final firms that they have bought and integrated into *gruppi* should seek to restructure subcontracting relations using contracts rather than informal cooperation. A bargaining account, in contrast, would suggest that large leader firms should behave in systematically different ways from *both* the smaller final firms that they have bought *and* the smaller final firms that they have not bought, while smaller final firms *in general* should behave in similar ways to each other, regardless of their broader ownership structures. Thus, the two accounts offer directly conflicting predictions regarding the behavior of small final firms which are members of *gruppi*.

Finally, a bargaining account would suggest that broad-scale institutional change (i.e., the collapse or even the substantial modification of the existing informal rules governing subcontracting relations in industrial districts) will only happen when actors in the community have good reason to believe that final firms (as a general class) will start to deal with subcontractors (as a general class) differently, and/or vice versa, in relevant situations. This will only occur if and when final firms (or subcontractors) start to behave differently *as a general class of actors* and succeed in repeating their actions over time against any opposition that might arise. Given the characteristics of this class, there is no good reason to expect that changes in the behavior of a small and idiosyncratic group of actors (very large firms with substantial internal production facilities) who belong to a relevant class of actors (final firms) will lead to changes in expected behavior on the part of members of the class of final firms as a whole. Because of their idiosyncrasies, very large final firms are both more likely than other firms to experiment with behavior that deviates from existing informal rules and less likely than other firms to precipitate broader institutional change. Hence, we will not see a change in the broad institutions governing subcontracting relations in industrial districts, and the forms of trust and cooperation that rely upon them, unless there are quite substantial changes in the opportunities open to final firms (or subcontractors) as a general class, rather than to a small group within that class.

It is more difficult to test these predictions against those of ownership structure accounts. On the one hand, the latter set of accounts typically do not seek to explain variation in outcomes, instead they often depict changes in ownership structures as a manifestation of all-encompassing secular forces of change, such as globalization (Harrison 1994) or modernization (Franchi and Rieser 1991). On the other hand, they are typically imprecise about the time frame within which change is supposed to take place, limiting their predictive capacity and their testability. Nonetheless, to the extent that one can show that the changes that these scholars have predicted have not taken place, and that the mechanisms that they invoke appear not to work as they have argued, one can reasonably state that their case is at best unproven, and at worst highly unlikely to be true.

5.3 CHANGE IN THE BOLOGNA PACKAGING MACHINERY DISTRICT AND ELSEWHERE

Which of the above accounts – the one that emphasizes ownership structures or the one that emphasizes informal institutions – best captures empirical changes in cooperative practices within industrial districts? In this section, I use evidence from the packaging machinery district in Bologna and from other Italian industrial districts to assess their relative merits.

Like many other industrial districts, the Bolognese packaging machinery district is clearly undergoing important organizational changes as a result of changes in demand markets. The district has gone through two main historical phases – one of expansion into open market, lasting until the early 1990s, and one of adaptation to changes in these markets, lasting from the early 1990s to the present, where large firms have come to play a more dominant role. This evidence supports the claims of authors who argue that large firms are playing a greater role and creating more hierarchical networks of subservient subcontractors in industrial districts such as Bologna. However, the evidence suggests that this increase in hierarchy has *not* led to the general collapse of informal practices of cooperation that these authors suggested was likely.

The Bologna packaging machinery industry has relatively shallow historical roots – before the World War II, there were only two firms in Bologna producing packaging machines: Acma and Sasib.[7] A third firm, GD, specialized in parts for bicycles and motorcycles before the war and machine-gun parts and military engines during it; after the war, it was faced with a drying up of military orders and moved into

[7] The following history of the packaging machine cluster is based on the accounts in Capecchi (1997), Curti and Grandi (1997), and Boari (2001).

the sector of automatic packaging machines, with the help of one of Acma's technicians (Capecchi 1997).[8]

These three firms were to be "mother firms" after the war, as technicians who worked for them moved on to found their own small firms. This was true of Acma in particular: at least ten former Acma technicians went on to form firms that later became important in the local industry.[9] It was relatively easy for individuals with technical skills to find new market niches – the mechanical techniques used to build packaging machines in one market segment could be transferred without difficulty to new products, and international markets for packaging machines were expanding rapidly.[10] However, it is also important to note that Acma, Sasib, and GD, which grew along with the rest of the industry, adopted a somewhat different pattern of production than the smaller firms that they gave birth to. They never relied on subcontracting as extensively as did other smaller firms: while they put out much of their work, they also retained substantial in-house engineering facilities.

The packaging industry in Bologna saw rapid expansion during the 1950s and 1960s, producing machines to package food products, pharmaceuticals, cosmetics, chemicals, and a plethora of smaller niche products. The *autunno caldo* of industrial unrest at the end of the 1960s was responsible for a major shift in production in the industry, as final firms began to outsource ever more of their work to subcontractors. This resulted in an enormous expansion in the number of smaller firms in the industry during the 1970s (see Fig. 5.1).

These subcontractors were typically artisanal producers with only a few employees. In the 1971 ISTAT Industrial Census, 113 mechanical engineering firms in the district had between 1 and 15 employees. By 1981, this number had more than tripled to 386 firms. While the number of very small firms has fallen in the succeeding two decades, it is still very high in relative terms. Most subcontractors remain quite small, although a few now have 50 or more technicians working for

8 GD's reliance on Acma's expertise in its early history in the packaging industry was later to be reversed; GD bought Acma in the 1980s from its then (U.S.) owners, to form a single group, within which, however, each constituent firm remained substantially autonomous.

9 Capecchi 1997.

10 Ibid.

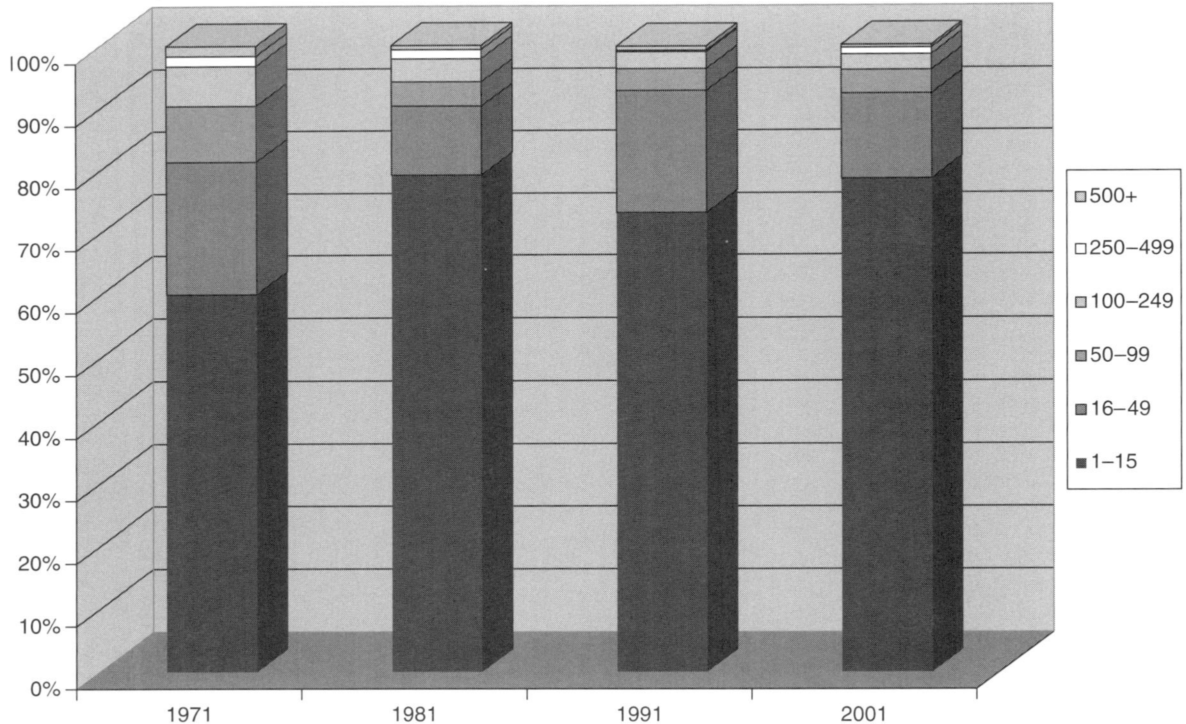

FIGURE 5.1. Distribution of firms by size category over time in the mechanical engineering industry in the province of Bologna. Vertical axis = number of firms, horizontal axis = year, shaded areas = size of firm in terms of number of people employed. *Source:* ISTAT Industrial Census data.

them. Although the number of larger firms also increased, it increased much less dramatically.

However, the underlying pattern of relationships between firms has changed rapidly since the early 1990s, largely as a result of change in demand markets. As in many other industrial districts, large "group" firms are emerging as an important and perhaps dominant form of organization. Using information from a shareholder database, Brioschi, Brioschi, and Cainelli (2002) estimate that 42.8% of relevant firms in the Bolognese packaging machinery district now belong to a larger group.[11] These groups seek to provide integrated ranges of machines within a particular sector, and sometimes in more than one sector.[12]

This process of integration was originally driven by a search for economies of scale. It appears now to be more driven by demand markets. Globalization has led to a winnowing out of the major product markets (food, pharmaceuticals) that packaging machinery vendors sell to. Many smaller customer firms have either disappeared or been bought out, leaving the field to very large multinational firms. These large customer firms prefer to deal with a single, integrated supplier, rather than with a variety of supplier firms. Previously, a customer might deal with a variety of different specialized firms, each of which produced one specific kind of machine for a production line. For example, a customer firm in the pharmaceutical industry might previously have bought a machine for dosing gelatine capsules with drugs from one mechanical engineering firm, a machine for packaging capsules into blister packs from another, and machines for boxing the blister packs and for palletizing cartons from other manufacturers still. Now, major customers tend to prefer to deal with only one firm, a preference that has led to a substantial shift of power toward larger firms, and an ongoing process of horizontal integration. In the words of one interviewee, summing up a consensus view shared by several other respondents:

[11] Note, however, that their data almost certainly overselects on medium-to-large-sized firms.

[12] The motivation for making packaging machines across more than one sector is to stabilize future income by making it less dependent on the changing fortunes of customers within a particular sector (such as pharmaceuticals or foods).

> the market reacts better certainly to a well-integrated range of products, rather than to a main supplier with subsuppliers. It is not seen as reliable – the supply done by your main supplier with subsuppliers.[13]

Previously, a large final firm manufacturing packaging machinery might work together with several smaller firms, which produced specialized machines that complemented its range. This is becoming less and less common as such firms are either being bought by larger firms or finding their niches disappearing. This is all the more marked because of changes in customer requirements. Increasingly, large customer firms do not ask simply for machines but for a broad array of services relating to these machines, of a sort which smaller firms would have difficulty in producing. Some large firms in the area are seeking to reinvent themselves and lock in relations with customers even more tightly, providing them with a complete "turnkey" service and perhaps eventually actually taking over parts of these customers' manufacturing processes for them. Relationships between machine producers and their final customers appear to be getting closer and closer, with knock-on consequences for large firm strategies:

> The [customer] firms, in pharmaceuticals, food etc., do not want anymore to have engineering offices internally and thus high costs, and they turn outside to expert people who are able to do everything. Considering that this doesn't exist, slowly the bigger firms, which have the economic possibilities, tend to acquire on the market those machines that they lack, or the firms that they lack for their chain.[14]

And, in the words of one of the large firms which has been most active in seeking to buy smaller firms:

> a lot of companies are asking now just to have one supplier for the complete line of the complete factory. So they force you to create a pool of different suppliers in order to interface them with just one person, one face. So the reason for the concentration is also because our customers are pushing us to concentrate.[15]

These changes have had important implications for the relationship between large firms and smaller final firms on the one hand, and large

[13] Firm Interview 15. See also Firm Interview 11.

[14] Ibid., 14.

[15] Ibid., 11.

firms and their subcontractors on the other, just as scholars such as Harrison (1994) and Brioschi, Brioschi, and Cainelli (2002) would have predicted. Some highly successful small firms still produce for niches that are unattractive for one reason or another to larger firms, but many now face the choice of integrating their operations with larger firms (being bought out or becoming a dependent subcontractor) or going out of business. As the managing director of one of the largest firms in the district puts it:

> in this field we exist really, when we have the control of the final customer. In manufacturing of industrial or capital equipment like this, if you manufacture capital equipment you have got to have the grip on the customer yourself. What you could say is that there is a space for a small company, but they will supply mostly through the sales organisations of larger companies. That means they are nothing, they can be purchased easily, or destroyed or eliminated. Because the market here is not for this machine, it is for these customers, and if you don't have these customers, you don't exist. You have got to have these customers to exist.[16]

Thus, the relative equality that used to exist between final producers no longer applies. Larger "group firms" are increasingly able to block other final firms from gaining direct access to customers. This increasingly allows them to dictate terms to smaller final firms, absorbing them into groups, engaging in direct takeovers, or driving them from the marketplace altogether.

Large final firms are increasingly able to dictate terms to their subcontractors too. Subcontractors in general face difficult market conditions:

> It is obvious that the big firm has the ability to play around, because there are too many subcontractors] today for the needs of the market. If one makes a contract on the market, supply is greater than demand.[17]

The very largest firms in the district have leveraged their power to move from traditional informal arrangements to formal and arm's-length contracts (Harrison 1994). In part, they have done this through making subcontractor relations quasi-hierarchical. In contrast to traditional practice in the district, where final firms interact directly

[16] Ibid., 15.
[17] Interview with Confartigianato official.

with a large number of subcontracting firms, the larger firms of the district are now increasingly seeking to develop a small number of subcontractors as "subsuppliers." These subsuppliers manage subcontracting relations on behalf of the final firm so that the large firm, like Japanese manufacturers, is at the apex of a vertically oriented "Christmas tree" of supplier relations or what Crouch and Trigilia (2001) describe as a "networked firm."

Clearly, this system still allows some forms of cooperation between subcontractors and very large final firms to take place. Equally clearly, these forms of cooperation are much less egalitarian, and they involve greater distributional gains for the buyer firm than under the previous dispensation. Indeed, the creation of this more vertically integrated system is conditioned on changes in the options available to different firms. The lock big firms have on their final customers allows them to be sure that they are not building up potential competitors by focusing their attentions on subsuppliers. The previously quoted managing director:

> Who could be possibly scared of a subsupplier acquiring some competence if I hold the customers? After all, I have the customers. He may learn to make a machine; he does not know how to conceive a machine, he cannot design a machine himself. He may copy my machine; and then to whom shall he sell it?[18]

Another large firm in the district has sought to develop a few of its subcontractors to a higher technical standard, helping them in the purchase of new machinery.[19] However, this has had considerable costs for the subcontractors in question. While they are highly dependent on the buyer firm, which accounts for around 80% of their turnover, they have to sign up to formal contractual terms that are "very favorable" to the buyer firm and have no guarantee of future work or income. As a senior manager in the large firm:

> We ask for a lot of flexibility from our suppliers! That is the main concern they have normally. They don't like to be treated in that way, because for us it is difficult to predict what is the workload that we would pass to our suppliers. So normally our projection is always pretty wrong. But on the other side, for them we are very important. So they just complain.[20]

[18] Firm Interview 15.

[19] Ibid., 11.

[20] Ibid.

This is quite different from the traditional dispensation in which both final firm and subcontractor shared risks. Subcontractors formerly independent partners in the production process are now becoming "flexibility reservoirs" (Semlinger 1993b) for larger ones.

Thus, as industrial district skeptics would have predicted, the impetus toward the creation of "groups" of firms has been associated with a very substantial increase in hierarchical relations (1) between very large final firms and other final firms, and (2) between very large final firms and their subcontractors. These changes in power relations have gone together with much greater use of contracts between these large final firms and their subcontractors, effectively replacing informal norms of mutual honesty and flexibility with one-sided formal agreements.

However, as the evidence from Chapter 4 demonstrates, these increases in hierarchy have *not* led to the general destabilization of informal practices of cooperation that authors like Harrison have predicted. Most importantly, the evidence clearly indicates that while there have been important changes in the relationship between very large final firms and their subcontractors, these changes have been disseminated neither to smaller final firms within *gruppi* owned by larger firms nor indeed to independent final firms.

Alongside hierarchical relations of exploitation involving a small number of very large firms and their subcontractors, the larger population of smaller final firms continues to use informal agreements that instantiate relatively egalitarian power relationships with their subcontractors. These final firms show no signs of changing their subcontracting practices to bring them in line with those of larger firms. As the interview evidence discussed in the previous chapter makes clear, smaller firms rely extensively on existing informal rules dictating honesty and fair dealing as a means of anchoring broader relationships of gift exchange.

This is as true of smaller and mid-sized final firms that have been bought out and are now part of larger *gruppi* as it is of smaller final firms that have remained independent. The evidence suggests that smaller firms in *gruppi*, like their independent counterparts, rely on diffuse oral agreements with their subcontractors and extensive forms of cooperation in which the boundaries between the firm and its subcontractors were sometimes hard to draw[21]:

[21] See, for example, Firm Interview 13.

> The subcontractors are of critical importance, because when we delegate the making of pieces outside, it is as if we put a half, a part of the firm outside.[22]
>
> As I said already, we use many suppliers, both for mechanical parts, and electronic parts. We do the assembly, the planning of the machine ... Then, at a certain time, we give designs to the various suppliers which we have – local suppliers, already there for several years. This can even be understood as the "periphery" of [firm name].[23]

In short, these final firms have not abandoned traditional modes of cooperation based on informal community rules. Instead, they have continued to put out most of their work to subcontractors and to rely on close relationships of trust and cooperation with these subcontractors in so doing. Their logic of organization means that they are bound up in relations of mutual dependency with these subcontractors, and they thus need to continue to maintain close and mutually trusting relations with them in order to be able to produce machines successfully.

There is some evidence that larger firms initially failed to understand this (Harrison 1994). When larger firms began to assemble *gruppi* of smaller final firms under their aegis, they indeed tried to impose common sourcing structures upon these firms in order to achieve efficiencies of scale. However, as interviews with larger firms revealed, this strategy proved to be unworkable and was swiftly abandoned. Group firms in the district currently share few decision-making structures beyond a common financial structure and shared arrangements for sales;[24] they certainly do not centralize decisions over subcontracting. The managing director of one of the largest firms in the district:

> The approach we choose for the rest of the business, is to create local business units which are completely independent, and they must respond to their results. They are completely independent, nobody from here can interfere with their business. ... Where they take advantage of the critical mass is on the finance side. ... Second, on the [sales] ... There are a lot of savings that you can do. But they are still independent, that is important. Because we tried also to unify the organisation in the past, but it was a complete disaster. Because you lose control over the focus of your business ... they don't have to

[22] Firm Interview 7.

[23] Ibid., 5.

[24] Ibid., 15, 18.

buy the same components, they don't have to use a certain subsupplier, they are completely free.[25]

Subcontractors still face more difficult market conditions in the past because some smaller final firms that have been bought up have been directly integrated into larger ones rather than incorporated into loose group structures, so that subcontractors have fewer potential customers. However, one cannot simply read from market conditions to changes in power relations. The tendency of very large firms to press for more hierarchical relations with subcontractors does *not* mean, as authors such as Harrison predicted, that these practices are diffusing to smaller final firms within the larger firms' *gruppi*, let alone within the district more generally. Instead, smaller final firms that have been bought up by larger ones continue to adhere to localized informal norms of mutual fair dealing, rather than moving toward formal contracts and associated arrangements that would lead to a more asymmetric distribution of the benefits of cooperation (see also Lazerson and Lorenzoni 1999).

This suggests, contrary to the predictions of scholars who focus on ownership structures as a key vector of change in industrial districts, that small final firms within *gruppi* behave like their independent counterparts rather than like the large final firms that bought them out. Very large final firms, as bargaining theory would predict, have been far more willing than small firms to experiment with new forms of production, which directly challenge existing informal institutions. Large final firms have also been far more willing to maintain these forms of production in the face of unhappiness from their subcontractors. Some smaller final firms bought out by larger ones did indeed temporarily adopt pooled subcontracting arrangements under pressure from their new owners, but they swiftly abandoned these arrangements when they proved unworkable.

Even though the interviews do not provide direct evidence that these arrangements were abandoned *because* of their consequences for informal relationships with subcontractors, the eagerness with which interviewees reembraced traditional diffuse arrangements of cooperation suggests that this is, at the very least, a plausible partial

[25] Ibid., 15.

explanation of their actions. Thus, the available evidence appears to support an institutionalist interpretation in which actors' willingness to take and sustain actions that are at odds with existing informal rules varies with their bargaining strength, as measured by their degree of dependence on other actors. Large firms, which were not dependent on their subcontractors, proved much more willing and able to take deviating actions and to sustain them over time, than smaller final firms.

By the same token, the deviating actions taken by very large firms have not seriously challenged existing informal institutions, precisely because these actions cannot be generalized across the group of final firms as a whole. Thus, instead of institutional change or breakdown, we have seen the withdrawal of a small (albeit important) group of actors from the existing system of rules, without any very substantial change in how those rules operate for the large majority of firms. Informal institutional change requires a much broader set of changed behaviors across the members of the relevant class of actors, an acceptance on the part of other actors that this is indeed the expected behavior for members of that class of actors going forward and, perhaps, appropriate changes in expected strategies on the part of other classes of actors in response to the changes that have been pushed through. These conditions were not satisfied in this instance.

Evidence from other industrial districts suggests that similar mechanisms explain change and stability there too. Josh Whitford (2001b), in his study of shoe producers in the Marche district, finds that norms of informal cooperation have survived both difficult market conditions and an increase in hierarchical relations among firms. A few key firms with well-known international brands are prospering while other firms are increasingly relegated to niche markets or to subcontracting for their more successful peers. However, even though bargaining strength has shifted decisively to firms with a direct presence on consumer markets, these firms have good reason not to squeeze their subcontractors too much, as the latter have specialized knowledge that is difficult to replace. While final firms are able to reap the benefits of their more powerful bargaining position, they have been careful not to take maximal advantage of their increased bargaining strength, so that existing informal practices of fair mutual dealing continue to survive:

> A firm that informally guarantees work to 10 factories (without contractually binding themselves) recognizes its superior bargaining position, but tries to keep long-term relationships because "there is collaboration. Today the market is made up of firms who collaborate with each other. At the level of mentality, we do not have an interest in taking advantage. Our interest is in having collaborators who are able to give us a service, a product, a good price." (13)

In return for their forbearance they receive a degree of informal cooperation that they would be unlikely to receive if they squeezed these subcontractors harder. Thus, existing informal institutions appear to have survived. Similarly, Sebastiano Brusco and Daniela Bigarelli (2002) find that subcontractors in knitwear districts in Emilia-Romagna are less likely to be squeezed for profits by final firms than one might expect, because final firms would have considerable difficulty in finding replacement subcontractors if the relationship broke down, while "if [the final firm] does not behave correctly, the subcontractor can readily find another customer" (Brusco and Bigarelli 2002, 27).[26] They argue that this continuation of cooperation is linked to the continued existence of a "complicated system of rules that includes a highly structured, although unwritten, mixture of trust, inspections and sanctions" (28).

In both these cases, informal institutions seem to have survived the advent of an increase in hierarchical relations among some firms. Gabi dei Ottati (2003) reports a more troubling set of outcomes among textile producers in the much-studied industrial district of Prato. Carlo Trigilia (1989) has argued that in an earlier era there were only

[26] Brusco and Bigarelli also claim that informal institutions affect the extent to which final firms are able to take advantage of subcontractors by arguing that "final and subcontractor enterprises acquiesce in accepting a climate of substantial cooperation and agree to coexist in a complicated system of rules that includes a highly structured, although unwritten, mixture of trust, inspections and sanctions. The impact that the degree of dependency has on subcontractors' profit levels is determined by the existence and efficiency of this system of rules" (28). However, Whitford (2001b) finds no evidence that rules have this impact in his empirical case study. The theoretical framework that I propose in this chapter would suggest that Brusco and Bigarelli have the causation backwards – it is the degree of mutual dependency that shapes rules supporting trust, inspections and sanctions, rather than vice versa.

moderate asymmetries in bargaining power between Prato's *impannatori* (roughly speaking, final firms) and subcontractors, and hence in patterns of trust and cooperation:

> It is these [trust relations] that allow the rapidity of adaptation to the market. They favour the exchange of information and knowledge, reduce the costs of a complex formal contract, and attenuate potential opportunistic conflicts deriving from the opportunistic taking advantage of temporary asymmetries in market relations (even if it is obvious that the firms making the order maintain a position of greater contractual power). In practice, thus, trust-relationships tend to stabilise market relations between firms. (Trigilia 1989, 294–95, my translation)

However, asymmetries in the trust relationship seem to have been greatly exacerbated in the wake of both globalization and increased competition from producers in China and elsewhere. Gabi dei Ottati (2003) finds strong evidence for the existence of informal local norms of conduct governing fair dealing – some 87% of respondents in a survey of Pratese business figures confirmed the existence of such norms. Nonetheless, their effectiveness is coming under challenge; many of dei Ottati's interviewees suggested that traditional means of informal sanctioning were no longer sufficient to underpin honesty in commercial relations. Furthermore, while only 5% of business actors in final firms (*impannatori*) believed that they needed to fear opportunism from subcontractors, nearly 60% of subcontractors reported that they feared that final firms will behave opportunistically when given a chance. Dei Ottati attributes these changes both to increased market pressures from China and elsewhere and to an increase in the bargaining strength of the purchasers of textile products that has created vastly more difficult conditions for all *impannatori*. Her diagnosis is worth quoting at length:

> These external changes, combined with the international slump, seem to have reduced the market power of many district final firms to the point of triggering cut-throat competition among them. Considering the way in which production in the district is organized, all this has brought about a progressive reduction in the rates paid to local subcontractors, an extension of terms of payment and an increase in objections made by external clients to the Prato final firms, and from the latter to local subcontractors. No wonder, then, if this situation tends to erode trust between Prato firms, in particular the subcontractors' trust in their buyer firms, and above all, the trust based on

respect for the local code of fair competition, which is distinctive of the district form of organization. (dei Ottati 2003, 10)

Here, unlike in Bologna, the Marche, and the knitwear districts of Prato, the conditions for informal institutional change appear to have been met. The spur for change here is not opportunity but a roughly symmetrical crisis, which has affected all *impannatori*. All of these firms face vastly more difficult market conditions than in the past, not only because of more stringent demands from final customers but also because of enormous pressure from Chinese competitor firms in the wake of the winding down of the General Agreement on Trade and Tariffs (GATT) Multifibre Arrangement, which had previously protected textile producers in developed countries such as Italy from potential competitors in the developing world. This means that *impannatori* face similar – and highly salient – pressures, which have led them to adopt new and less scrupulous tactics vis-à-vis their suppliers (such as frequent spurious objections to the quality of products) than in the past, despite previously existing norms of fair dealing. This is in turn leading a substantial majority of subcontracting firms to expect dishonest behavior on the part of final firms.

In Prato, external pressures are forcing final firms as a collectivity to behave opportunistically in ways that may damage long-term relationships in order to ensure their short-term survival. Under these circumstances, one may reasonably claim that institutional change is indeed occurring. A majority of subcontractors now fear unscrupulous behavior on the part of *impannatori* in general, inferring that specific acts of dishonesty on the part of particular final firms form part of a broader pattern of behavior on the part of all final firms. When actors of class *A* (subcontractors) encounter actors of class *B* (final firms) in situations where existing informal institutions previously mandated honest behavior, they can no longer expect these institutions to apply as before. Hence, the scope of these informal institutions is being redefined and narrowed. Specific battles over whether this or that behavior is appropriate are now being fought on terms that favor the *impannatori*, leading to a redefinition of the scope of those institutions, so that forms of behavior by final firms that would previously have incurred sanctions are now reluctantly accepted by subcontractors.

Thus, at the very least, this is leading to a substantial redefinition of the informal institutions governing relations between final firms and subcontractors such that subcontractors may no longer trust final firms over important transactions.

5.4 CONCLUSIONS

It is clear from the empirical evidence that the Bologna packaging machinery district is changing, just as other Italian districts are. First, relations *between* final firms are changing in some very important ways. As a result of changes on international demand markets, smaller final firms are being bought up by larger ones and assembled into *gruppi* in order to present complete product lines to large international customers. Second, relations between a very small number of final firms (the very largest firms in the district) and their subcontractors are becoming more hierarchical and formalized – and less reliant on trust relations.

However, it is equally clear that these changes are not having the dire consequences for broader patterns of informal trust and cooperation between final firms and subcontractors that many expected. This is especially notable given that Bologna has previously been presented (Harrison 1994) as a harbinger of the likely unsustainability of the Italian industrial district model. Contrary to the predictions of industrial district skeptics, business practices based on informal trust and cooperation appear to be flourishing.

In this chapter, I have sought to explain this by isolating the possible causal pathways that might better explain when we can expect stability and when change. In a first step, I have tested two causal accounts – one privileging ownership patterns as the vector of causation, the other privileging informal institutions – against each other. The former is derived from the claims of the major strand of skepticism over the viability of industrial districts, the latter weds a theory of institutional stability and change to the arguments of those who believe that informal rules have played a vital role in explaining outcomes in these districts. Both accounts make similar predictions regarding the actions of one group of actors – very large final firms. Although they make dissimilar predictions with regard to small independent final firms, it is difficult to use this particular subset of predictions to test these two accounts against each other; the ownership structure approach is notably vague as to the precise mechanisms through which changes in informal practices are propagated to final firms, and thus offer little insight into the timing of change. Scholars who privilege ownership structures in their explanations might reasonably claim that independent final firms will eventually conform to their predictions but that this will take a long time.

They can make no such arguments with regard to small final firms that have been bought out and integrated into *gruppi*. Under ownership-structure accounts, these smaller firms should adopt – and stick to – the subcontracting practices of their new owners, replacing practices of diffuse cooperation based on word-of-mouth agreements with formalized arrangements at the group level that take advantage of economies of scale. An institutionalist approach of the kind that I have presented would predict, in contrast, that these firms will continue to rely on practices of diffuse cooperation, just like independent final firms, since they continue to be highly dependent on their subcontractors and hence unlikely to deviate from existing informal institutions except in response to truly significant opportunities or threats. Thus, small final firms that have been bought out are "critical cases" that allow us to test the one theoretical account against the other.

That these critical cases support the predictions of the institutionalist account rather than those of the ownership-structure account suggests that the former provides us with a better understanding of the relevant causal mechanisms than the latter. Clearly, large-firm strategies have an important (and likely an increasingly important) role in industrial districts; to take just one example, they have reshaped relations among final firms in very significant ways. Equally clearly, they have not had the broad consequences for trust and cooperation between most final firms and subcontractors that ownership structure accounts predict.

In a second step, I have argued that informal institutional change will only happen when three criteria are fulfilled – (1) that deviation from existing informal rules benefits initiating actors, (2) that they are able to sustain that deviation against opposition from other actors who might prefer the status quo, and (3) that deviating actions can be copied by all or most others in the relevant class of actors, so that it becomes the expected response from members of that class in situations of a given sort. The detailed account of stability and change in Bologna would appear to bear out this argument.

First, deviating actions (the adoption of more formalized subcontractor relations) were undertaken by a very small group of final firms that were much less deeply embedded in reciprocal relations of dependency than most other final firms. Second, these firms, when they found that these actions advantageous were able to sustain them, since the actors negatively affected by them (their subcontractors) had relatively little bargaining power to resist them. Third, when these firms tried to get other actors – smaller final firms that they had controlling interests in – to take up these actions too through formalized group-level structures, they found that this was, in the words of one of the large firms involved, a "disaster" and was quickly abandoned by smaller final firms within these groups in favor of a return to traditional practices. Finally, these actions never disseminated to the vast majority of other independent final firms, so that they never led to broader changes in expectations over how classes of actors interact, and hence they never led to change in the relevant informal institutions.

Here we can see how, as Hall and Thelen (2007) and many economic sociologists suggest, institutional stability and change involves actions that test and retest boundaries, sometimes succeeding, sometimes failing. In general, rational-choice accounts of institutional change have overlooked microlevel processes of this kind, providing instead more abstract accounts of how change occurs. However, as I have argued here, exactly such microprocesses are necessary to explain when the actions of particular actors lead to changed expectations over how general classes of actors should be expected to behave, and hence institutional change, and when actions do not have these consequences. Sometimes, actors may make mistakes, either over their interests or their ability to sustain actions against opposition. Large firms in Bologna were clearly mistaken in their efforts to make the smaller final firms in their groups conform to new subcontracting practices, and they had to abandon these efforts. Sometimes, actions may not disseminate across the class of actors in question because most actors in the relevant class are unable or unwilling to taken them up. Under either of these circumstances, actions will not precipitate change in informal institutions.

This account suggests that the informal institutions governing subcontracting relations in industrial districts should be considerably more stable in the face of changing global markets than skeptics have suggested. However, it also suggests that we will see change in informal institutions where there are genuinely substantial changes in the outside options available to actors that overwhelm the advantages of existing relations under informal institutions. Such changes might reasonably involve either substantial new opportunities that provide "attractive" outside options to actors belonging to relevant classes or threats that make the existing status quo much less attractive in relative terms than it had hitherto been. The textile industry in Prato provides a cogent example of how genuinely serious threats to the status quo can lead to quite significant changes in informal institutions.

Finally, by providing a concrete analysis of the mechanisms of change and stability in informal institutions, this chapter makes a significant contribution to current debates over political economy. It provides a basis not only for arguing that informal institutions are unlikely to disappear in the ways that some globalization accounts suggest but also for distinguishing between situations where informal institutions are likely to remain stable and those situations where they are likely to come under significant challenge.

Nonetheless, even if industrial districts provide an unusually crisp example of the role of informal institutions in advanced industrial economies, they are, by the same token, cases that are difficult to generalize from. The pathway identified here is unlikely to be the only chain of causation through which changes in global markets may affect the governance of domestic or local economies. For example, Pepper Culpepper (2005) provides a convincing account of informal institutional change and financial institutions, in which ideational factors play a determinative role. In the absence of some overarching master account, the claims made in this chapter should be taken less as a universal account of how informal institutional change occurs than an attempt to identify a specific set of mechanisms, among a broader universe of mechanisms, that may lead to informal institutional change. It is highly likely that this pathway will be replicated in other settings; however, it will almost certainly not be the only set of mechanisms through which informal institutional change may occur.

A more satisfying account of informal institutional change would ideally incorporate *both* ideational *and* bargaining power factors. However, despite continuing efforts by a variety of scholars, we are still very far from articulating such a theory. If such an account is indeed possible, it is reasonable to suspect that the best way to build it is precisely through articulating more precisely the microprocesses of informal institutional change and seeking to identify the ways in which different causal mechanisms work and (where more than one such mechanism is present) how they interact. This involves significant challenges both for rational choice accounts (which often tend to abstract away the details of institutional change) and sociological approaches (which frequently forsake broader theory building in favor of concrete analyses of specific contexts).

However, even if it is an insufficient basis for a proper theory of how ideas and material factors intersect, evidence from Bologna and other districts does suggest some important limits to the role of new ideas in changing collective beliefs. The large firms that bought up other final firms in the packaging machinery district were central actors in the district (all other firms were highly aware of their activities; most small firms had been founded by people who formerly worked for them). Under some ideationalist arguments (e.g., Culpepper 2005), one might reasonably have expected their ideas regarding the organization of production to have led to large scale change in the district. Instead, when they sought to radically reorganize production, they succeeded only in reshaping their own supply relationships, not those of the smaller final firms that they had bought out nor those of the district as a whole. Their ideas failed because of important differences in the material conditions of the various final firms in the district.

This is not to say that ideas are irrelevant to institutional change – far from it. One might imagine that under other circumstances, different ideas might have succeeded in reshaping the district (as other ideas have in other environments). That is to say that we still lack a proper account of the circumstances under which ideas propagate and under which they do not. A better understanding of the circumstances under which ideas do or do not have these consequences will likely require scholars of institutions to go beyond both standard constructivist/ideational and rational choice accounts. One highly promising research direction is the emerging body of research in sociology, epidemiology, and related fields on heuristics, propagation, and networks (Page 2007; Bednar and Page 2007; Watts and Strogatz 2006). Building on this agenda, however, would go considerably beyond the task of this book.

6

Informal Institutions without Trust

Relations among Mafiosi in Sicily

6.1 INTRODUCTION

In Chapter 4, I argued that there are important differences between economies (such as the industrial districts of central and north-eastern Italy) where informal institutions play an important role, and economies (such as the machine producing industry of Baden-Württemberg) where formal institutions dominate. I suggested that we may expect trust relationships to play a much more important role in societies where informal institutions predominate, while more specific expectations over institutional compliance will anchor important cooperative relations in societies with strong formal institutions.

This argument, if left unqualified, might by omission suggest a kind of functionalism that is quite as specious and unfaithful to empirical reality as the economistic forms of functionalism discussed in Chapter 2. To say that trusting relations are highly important in societies dominated by informal institutions is *not* to say that societies dominated by informal institutions will necessarily support rich and wide-reaching forms of trust. Indeed, the available empirical evidence makes it emphatically clear that informal institutions need not be associated with trust. Informal institutions play a dominant role in many societies in the global South that would appear to have low levels of interpersonal trust and cooperation, whether measured through the imperfect gauge of survey evidence (Nannestad 2008) or direct observation of actors' behavior.

Thus, the claim that informal institutions can support forms of trust that formal institutions cannot is no more than a possibility claim. One may still plausibly argue that trust will play a highly important role in societies where informal institutions rather than formal institutions structure economic exchange. In the absence of formal institutions that might underpin impersonal exchange, individuals will often have to fall back on trust relations when they need to engage in complex and risky transactions. However, there is *no guarantee whatsoever* that such trust relations will exist for many kinds of transactions, or that if they do exist, they will be at all extensive. Edward Banfield's famous (1958) and controversial (Sabetti 2000) work on amoral familism in southern Italy describes a context in which trust relations are both enormously important and hard to find outside the immediate family. Here, trust is important because the formal institutions that would underpin impersonal exchange are grossly underdeveloped, but it is unlikely to develop between people who are not closely related to each other.

Thus, one should not generalize from relatively high-trust informal institutional settings like that discussed in the previous chapter to the claim that all informal institutional settings are likely to produce high trust. The empirical evidence strongly suggests that some informal institutional settings are characterized by pervasive and generalized mistrust over important areas of political, economic, and social exchange.

Indeed, certain informal institutions may have directly malignant consequences for trust. As Helmke and Levitsky (2004) argue, corruption may be understood as a consequence of certain kinds of informal institutions. Generalized expectations that, say, ordinary citizens should pay a bribe to police officers if they are pulled over to the side of a road are informal institutions in precisely the sense that I have discussed in previous chapters of this book. They involve unwritten rules over how actors of certain types (citizens and police officers) should interact in situations of a particular kind, with some fuzziness around the edges. Farrell and Knight (unpublished) further argue that these dynamics help explain why highly corrupt societies are usually associated with high levels of general mistrust – they are societies in which the informal institutions do not provide a basis for equitable exchange between actors from different social categories, but instead give actors of certain social categories (ordinary citizens) justified reason to fear that actors of other categories (police officers or public officials) will behave in an untrustworthy fashion if they are not paid off.

How may we distinguish between contexts, such as Bologna, in which informal institutions produce extensive trust and cooperation among actors and other contexts dominated by informal institutions in which they most certainly do not? In this chapter, I build upon the arguments advanced in Chapter 2 to argue that the key difference between trust-supportive and trust-destructive informal institutions has to do with how institutions do or do not transmit information about others' intentions. I go on to apply these arguments to an important test case – internal relations within the Sicilian Mafia. I do not suggest that we can conduct a controlled comparison between the kinds of trust or distrust that we might observe within these organizations and the kinds of trust or distrust that we observe in industrial districts such as the packaging machinery district of Bologna, much less the machine tool producing cluster around Stuttgart. Very clearly, the matters over which actors trust or distrust each other are different across these cases. Independent of relevant institutions, we might reasonably expect mafiosi to be less likely to trust each other than Bolognese businessmen (the interests of the former will frequently clash in a far starker manner than the interests of the latter).

Instead, I seek to demonstrate that the relevant informal institutions are an independent source of distrust even apart from clashes of interests and, more precisely, to show how informal institutions that cast a malign penumbra over expectations regarding future situations can help to exacerbate and to perpetuate existing patterns of distrust among actors. By so doing, I establish that informal institutions can play an important causal role in provoking distrust as well as trust among actors.

These arguments build upon work in the existing literature, most obviously and importantly Diego Gambetta's landmark research on trust (1988) and the Sicilian Mafia (1993). Yet they also highlight factors and causal relationships that are underexplored in Gambetta's work. Gambetta's primary aim is to explain relationships between mafiosi and their clients and the consequences of these relationships for generalized patterns of distrust and trust in Sicily. He is especially interested in the role that signaling plays in shaping beliefs that may in turn affect actors' propensities to trust or to distrust others. My arguments examine related but often dissimilar questions of trust and distrust among mafiosi *themselves*, a topic that Gambetta provides rich qualitative information on, but does not explicitly theorize. They also highlight the role of informal institutions in dispelling or exacerbating the ambiguities over intentions that make signaling of intentions more or less important.

The testimony of former Sicilian mafiosi describes an informal rule (that "men of honor," or members of the Mafia, should always tell each other the truth), which appears at first glance to be similar to the informal institutions conducting toward honest behavior in Emilia-Romagna. However, this rule has quite different consequences for trust. This informal rule of honesty among mafiosi is not only imperfectly enforced but it is modified by other norms that have the effect of making mafiosi's ultimate intentions extremely difficult for each other to discern. This helps to create an environment of mutual distrust among mafiosi in which, as Gambetta notes, they monitor each other constantly for signs of their actual intentions. Furthermore, the relevant informal rules have created systematic blockages in information flows that can be (and are) manipulated by privileged actors within the Mafia in order to provoke mutual distrust among other mafiosi.

In developing these arguments, I use a variety of primary and secondary source material, drawing extensively on a database of primary source information collated from the confessions of Mafia *pentiti* and a variety of other police and judicial material that has recently been made available by Diego Gambetta (Gambetta's database is henceforth referred to as GD), as well as on other sources. The kind of information collated in this database has some obvious problems. We cannot be sure that repentant mafiosi are always telling the truth. In particular, there is strong reason to suspect that the wave of *pentiti* who began to come to the police in the late 1980s and early 1990s had reason to exaggerate the efficacy and universality of informal norms of honesty. Nonetheless, by comparing information from Mafia confessions, both with each other and with other sources of information, we can cautiously draw conclusions about the way in which informal institutions work within the Mafia (Stajano 1992). Moreover, as Gambetta drily observes, there is little alternative for scholars who are interested in the internal workings of Mafia-type organizations:

> Virtually no study of the mafia is based on fieldwork. Scholars do not like to waste their time with uncooperative sources who refuse to talk or, alternatively, to be shot; they have therefore wisely concentrated on historical and judicial sources. (1993, 9)

In the remainder of this chapter, I first lay out my theoretical arguments at greater length. Then, I examine the informal institutions that govern relations within the Sicilian Mafia. Next, I examine the consequences that these institutions have for trust and distrust among actors, drawing both on confessions from Mafia *pentiti* and from other sources of information. I conclude by examining the similarities and differences between the informal institutions governing relations among business actors in Bologna and Sicilian mafiosi in greater detail.

6.2 INFORMAL INSTITUTIONS IN THE MAFIA

Under what circumstances is it likely that informal institutions will lead to distrust rather than trust? To answer this question, it is necessary to return to the arguments canvassed in Chapter 2. As a general matter, informal institutions are important not because they provide precise expectations over well-defined situations, as formal institutions do, but because they cast a penumbra over a broad and imperfectly defined set of future situations. Actors may reasonably turn to these institutions for guidance as to how others are likely to behave in such situations. This is because informal institutions provide actors with information about others' intentions toward themselves in these situations, and most particularly about whether those others' intentions will lead them to behave in a trustworthy or untrustworthy fashion.

This information is of two kinds. First, there is information that derives from the institutions themselves. Institutions instantiate broadly held perceptions about the ways in which different classes of actors are likely to interact with each other. Thus, individuals belonging to these classes can draw on institutions for information about how other actors belonging to these classes are likely to act in particular situations and about how they themselves are expected to behave. Second, institutions shape the ways in which actors do, and do not, communicate with each other, often dictating which matters and actors are discussed, what kinds of information are disclosed about them, and so on. Since individuals may draw on this information in order to make inferences about the specific intentions, and thus the trustworthiness of others, the underlying institutions governing communication may have important direct effects for trust.

In situations such as those discussed in Chapter 4, we may expect that relevant informal institutions will frequently support trust because they provide actors with grounded reasons to believe that other actors will behave in trustworthy ways across a wide variety of situations. They lead the relevant actors to conclude that others' intentions toward them are such that those others will behave in a trustworthy fashion, and that if they themselves cooperate in appropriate ways, their cooperation will not be traduced. They also provide relatively swift and effective means through which information regarding the trustworthiness or untrustworthiness of particular others can be communicated within the relevant community.

However, other kinds of informal institutions may cast a less happy penumbra. Where the relevant informal institutions provide actors either (1) with grounded reason to believe that others will behave in an untrustworthy fashion or (2) with reason to suspect that others' intentions toward them are ambiguous and possibly malign, we may expect that these institutions will not support trust over situations that fall beneath their penumbra, and that they may indeed actively promote distrust in these situations. Here, informal institutions are likely to lead to distrust and untrustworthy behavior. Furthermore, where institutions place sharp constraints regarding which actors are allowed to communicate with which, and what they can discuss, we may expect that their ability to support trust and cooperation will be correspondingly limited, as they will be less well suited to providing information about others' behavior and likely intentions. We may also reasonably anticipate that institutionalized gaps in communication will readily be exploited by well-placed agents to manipulate the information that is available to others in ways that can further exacerbate problems of distrust.

This helps us understand how informal institutions affect trust and cooperation within the Mafia. There is good reason to believe relations among mafiosi are indeed shaped by informal rules and norms. Up until the 1990s, there was controversy among scholars over whether the Sicilian Mafia was best understood as a coherent organization or as a congeries of unlinked groups sharing similar practices (Catanzaro 1985). More recent scholarship suggests that the answer lies somewhere in between; even while the Mafia is not a single hierarchical organization, different Mafia families share many organizational structures in common, including, most importantly, a head of family, who is (at least in formal terms) elected by the "men of honor" within the family, and *capodecinas* who work under the head of family and are responsible for directly supervising a number of "soldiers."[1] In addition, the most important grouping of families within the Sicilian Mafia has had an overarching institution, the so-called provincial *commissione*, which comprises representatives from various families, so as to coordinate expectations and settle (or at least manage) cross-family disputes in the Palermo area. Initially, it appears that the *commissione* was intended to check the problem of intrafamily violence by providing a veto point that could approve or fail to approve murders within families; this proved, however, to be unworkable in practice.[2]

These shared structures are important but do not mean that the Mafia is a traditional hierarchical organization. Instead, it is structured around smaller organizations – the families themselves – which

[1] There are no women Mafia members; this stands in contrast to the less organized Neapolitan Camorra, where women have sometimes played an important leadership role.

[2] "This *commissione* was born exactly to block all of the horrible things that happened, in which everyone was the master of the life of everyone else ... And then every murder, even if it was by the family within the family, was to go to the *commissione*, which would say who lived and who died ... However, this didn't work for long" (Record 2868, GD, my translation). Buscetta's testimony supports these claims; he suggests that the *commissione* was originally set up to settle disputes between ordinary mafiosi and their chiefs, and was then extended to manage interfamily matters (quoted in Paoli 2000). Even while the *commissione* failed to maintain control of intra-family killings, it was supposed to be consulted before major murders (e.g., the killing of prominent government officials). Furthermore, the norm that no killings should take place on a family's territory without the permission of that family's boss appears to have been widely understood, if occasionally breached. In Buscetta's words, "I want to underline emphatically that no homicide can be committed in the zone of influence of a particular family, without the blessing of the head of that same family. Major homicides furthermore require the agreement of the *commissione*. We are dealing with norms that don't permit any exceptions" (Record 1167, GD, my translation). In another statement he clarifies that murders can take place in the territory of a given boss without his permission, but that this is perceived as a direct challenge to the boss's authority (Record 1104, GD, my translation).

alternately compete or collude with each other as opportunities dictate. Gambetta (1993) suggests that the Mafia may be understood as a kind of cartel in which Mafia families play the role of oligopolistic firms, seeking to organize the market for protection among themselves.[3] More generally, Rocco Sciarrone suggests that the Mafia can be thought of as a "social contract," consisting of a "system of rules that creates constraints and opportunities of action for the various actors" (1998, 40, 41, my translation). Under Sciarrone's account, actors within the Mafia have a minimal shared interest in ensuring the survival of the organization, which may lead them to respond together to collective threats, but it also allows them to compete vigorously and sometimes viciously with each other inside their shared context.

Sciarrone suggests that there is good empirical evidence that informal rules play a key role in shaping interactions within the Sicilian Mafia. As described by the most famous of the *pentiti*, Tommaso Buscetta:

> No-one will ever find written codes of professional ethics, but the rules of which I became aware when I became a man of honor are rigid and ironbound and universally accepted.... the chains that bind men of honor to each other are more solid and unbreakable than if they were written down in any document. In my opinion, one of the main errors in the war against the mafia was precisely to ignore this truth, which is known to every man of honor. (Entry 1059, GD, my translation)

According to Buscetta's description, these informal rules, like the rules governing subcontractor-contractor relations within industrial districts, are known to everyone in the relevant community (men of honor) and share the other features that Knight (1992) identifies as necessary conditions for informal institutions.

Buscetta and other *pentiti* discuss the content of these informal rules in some detail.[4] Their accounts of these rules appear to have been borne out in broad outline by the discovery of a document, the so-called "ten commandments of the perfect mafioso," among the

[3] Another plausible comparison might be with the international political system, where states seek to retain control of their particular territories while adhering to a minimal set of shared norms, and sometimes coordinating on matters of common interest through organized structures.

[4] See Gambetta (1993, 118–126) for a very useful summarization in English of the various norms governing Mafia behavior.

personal effects of Mafia boss Bernardo Provenzano in November 2007.[5] These "commandments" include informational rules: no mafioso should introduce himself to another mafioso except in the presence of a third, who can vouch for the mafioso's identity, and mafiosi should always tell the truth when asked what they know about something. Other rules seem intended, at least in part, to help prevent information from leaking outside the Mafia: mafiosi should not socialize with the police and should stay away from bars. They should not steal money from other mafiosi or from other families. Other rules still cover matters of availability (they should turn up for meetings and should be available at a moment's notice if the Mafia requires it, even if their wife is in labor), or sexual mores (they should not lust after other mafiosi wives and should treat their own wives with respect). Nearly all of these rules (the apparent exceptions being the rules covering timeliness and availability, which seem almost to have crept in from a corporate employer's motivational pamphlet) have previously been described in detail in the confessions of one or more *pentiti*.

Clearly, some of these rules only have marginal relevance to matters of trust and cooperation among mafiosi. Certain of the rules governing sexual behavior, for example, seem most plausibly to be connected to the belief that mafiosi should have a certain public image

[5] See "Ecco il decalogo del perfetto mafioso," *La Repubblica*, November 20, 2007, for the contents of these ten commandments, and "Nelle carte del boss la mappa di Cosa Nostra: C'è anche il decalogo del 'buon Mafioso,'" *La Repubblica*, November 20, 2007, for an account of the conditions under which these commandments came to light. The *pentito* Salvatore Contorno, a soldier in the Santa Maria del Gesu family, describes his initiation oath (which he took some 30 years previous to the discovery of the Provenzano document) as having been a "kind of ten commandments – don't look at the women of other men, and always tell the truth" (Record 2988, GD, my translation). However, that these commandments were written down gives the lie to Buscetta's claim that the rules would never be found in written form; it also may be a significant indication of the weakening of the Mafia's internal cohesion in recent years. Antonino Calderone, another *pentito*, argues that proper men of honor never write anything down, describing how in the wake of the murder of a certain Leonardo Galante, a note was found at his house with an organizational diagram of the Mafia: "It must be said then, that the very fact of the existence of this note confirms that Galante wasn't a man of honour, because none of the internal matters [*nulla in seno*] of the Cosa Nostra can be put in writing." Calderone then went on to state that while he wrote notes during his interrogations as *aides-memoires*, he took care to destroy the notes before the questioning began (Record 1966, GD, my translation).

in order to maintain their reputation (although, as Gambetta argues, they may also be intended to prevent confidential information leaking out through pillow talk). In any event, the application of these norms is rather spotty (Gambetta 1993); powerful figures in the Mafia have quite openly had extramarital relations without repercussion.[6] Other rules (e.g., those governing relations with the police) concern the relationship of mafiosi to the outside world rather than each other.

The most directly relevant informal institutions to trust and cooperation among mafiosi themselves, unsurprisingly, are those that govern the ways that they communicate with each other. Buscetta describes the norms governing honesty among mafiosi in greater detail when he justifies his claim that his knowledge of events that he did not witness is accurate:

> Whatever is recounted by a man of honor in the presence of two other men of honor must always be truthful. He who breaks this rule, given that he has the option of not speaking, is liable to the most serious penalties, and to losing his life. In this case, the man of honor who has broken this rule comes to be referred to as a "spreader of tragedies" [tragediaturi]. (Record 1176, GD, my translation)[7]

It is not only possible but probable that Buscetta exaggerates the universality of this norm. Buscetta was one of the losers in an internecine war where the victors, the *Corleonesi*, proved especially adept at lying in order to achieve their ends. He almost certainly exaggerates the universality of this norm (thus denigrating the *Corleonesi* who broke it) in order to justify his own breaking of the norms that one should

[6] Perhaps unsurprisingly, the norms appear to be applied less rigorously to powerful and influential men within the Mafia. As Calderone describes it, "The companion of Francesco Ferrera enjoyed a reputation as a woman of easy virtue; this was an obstacle to Ferrera's nomination to become a man of honor. However, when one is speaking of important and well-known personages, even these obstacles can be overcome, as seen, for example in the case of his brother, and of Giuseppe Madonia of Vallelunga and Luigi Ilardo. Il Madonia even married a woman of this type" (Record 2082, GD, my translation). Elsewhere, Calderone notes sardonically that the rules of the Mafia are like the laws of the state: they apply to nearly everyone. A more recent *pentito*, Giovanni Brusca (2000), notorious even among *Corleonesi* mafiosi for his viciousness, records that his father elevated him to a higher position within the Mafia on the excuse that other possible candidates were excluded for one or the other reason (money, womanizing) under Mafia rules.

[7] I am grateful to Tommaso Sciortino and Daniel Davies for information on the etymology of "tragediaturi."

never disclose the internal affairs of the Mafia to outsiders or speak to the police.[8]

Even so, there is strong independent evidence, including the Provenzano document discussed above, that the norm of truth telling exists, even if it is sometimes upheld more in the breach than in the observance. Several *pentiti* refer to it; for example, an anonymous informant claims that in the Mafia, "speaking falsehood, lying, is considered one of the gravest of failings" (Record 13, GD, my translation). More generally, the *pentito* Antonino Calderone claims that men of honor do not need to sign papers, since their word is sufficient in and of itself (Record 1966, GD).

However, there are important limits to this norm too. The first of these limits has to do with the *kinds* of honest communication that mafiosi are required to engage in. While the relevant informal rules require mafiosi to speak honestly when they do speak, they do not require that mafiosi speak rather than remain silent or that they use plain and actionable language when they do speak. As described by Salvatore Contorno:

> Another fundamental principle among men of honor is the obligation to speak the truth with other men of honor. This is neither an obligation to speak, nor to answer anything except questions from one's own bosses; on the other hand, it is necessary not to be curious, or to ask about things where one doesn't have an interest. However, when one speaks, one must always tell the truth, and a violation of this rule is unthinkable. (Record 2832 GD, my translation)

Moreover, he claims that

> in a world like ours, where the less one speaks the better, half a sentence suffices to indicate perfectly well whether one is a member or not. (Quoted in Gambetta 1993, 121; see Record 2854, GD, for full context)

Tomasso Buscetta concurs:

> Within the mafia, no-one will give you a blow-by-blow account of a crime; it is enough, and one should never ask more, that a person makes it clear, even

[8] For their part, the *Corleonesi* accuse the losers in the Mafia war of having been involved in shady relationships with the police even before they began to confess and suggest that their own attacks were provoked by attempts by mafiosi and senior *carabinieri* to trap them. See Lodato (1999). Their other claim – that the losers in the war were primarily interested in maintaining control of the heroin trade – is rather more plausible.

> through his silence [that he was] the author of a certain crime ... With us, a gesture, a look, a wink of the eye is enough to understand exactly what happened. (Quoted in Stille 1995, 115)

In short, business conversations between mafiosi often seem to run the somewhat limited gamut between the laconic and the Delphic. As the Palermo magistrates describe it, "the circulation of news is limited to essentials" (Stajano 1992, 44). Gambetta (1993, 121–22) gives two specific examples of the opaque and indirect ways in which mafiosi discuss their activities. When Michele Greco spoke of the reason why Gigino Pizzuto was executed, he remarked that a man who signs IOUs and defaults must settle them sooner or later. When asked about the disappearance of a Vicari family member, that family's representative to the *commissione* responded that "once in Misilmeri a sheep vanished, and nobody has heard of it since," indicating obliquely that the family had had a hand in the disappearance.

These forms of communication make it difficult and sometimes impossible to discern individuals' underlying beliefs and motivations. However, the inherent ambiguities of these forms of allusive communication do not mean that they are useless. As the Buscetta quote above indicates, they often suffice perfectly well to disseminate information that mafiosi consider vitally necessary to share and to permit coordination of strategies as necessary.

Gambetta (1993) argues that these tendencies toward truncated and indirect speech have much to do with the mafioso's desire to maintain his reputation as the kind of self-controlled man who is neither overly inquisitive nor inappropriately inclined to give overly elaborate answers. While this is surely important (the evidence that reputation plays a key role in the behavior of mafiosi is decisive; see also Arlacchi 1988), one may also see these norms of terseness and obliquity as a consequence of the strategic environment that mafiosi find themselves in.

What evidence we have from the confessions and descriptions of *pentiti* suggests that questions of motivation are murky, and that mafiosi devote considerable attention to trying to decipher other mafiosi's speech, signals, and actions so as to try to discern those mafiosi's underlying intentions (see next section). In this environment, strategic considerations would suggest that individual actors will typically prefer not to provide too much information, given the likely willingness of others to take advantage of this information in potentially quite unpleasant ways. They will be especially loath to answer questions that would provide others with information regarding their underlying goals and motivations and allow those others to anticipate their actions and circumvent them, or use this information to create traps.

However, individuals' ability to avoid disclosing information will depend to a great extent on their power relations with others. Hierarchical inferiors within a particular family will find it difficult to avoid answering questions – even potentially dangerous ones – that are posed by hierarchical superiors. Hierarchical superiors, for their part, will be able to avoid sharing information with their inferiors except when they wish to. Contorno suggests above that the only occasion on which mafiosi are required to provide answers is when they are being questioned by their hierarchical superiors. These hierarchical superiors for their part have no necessary obligation to divulge information to those beneath them.[9] As Buscetta describes it:

> The family head informs, when he does it [at all], only those members of the family whom he considers worthy of receiving his confidences, and only to the extent that it seems appropriate. To give you an example, it is necessary to point out that one never asks questions of one's interlocutor in relationships between men of honor because this is seen as a sign of a regrettable curiosity and can be interpreted in unfortunate ways. (Record 1251, GD)

Thus, an examination of the strategic situation that mafiosi find themselves in would lead us to predict what seems to be the actual outcome – a set of norms governing communication in which actors on the one hand tend toward ellipsis and taciturnity where possible, so as to avoid giving away potentially dangerous information, but on the other may find themselves obliged to answer inquiries from those who are in a superior hierarchical position to them. Buscetta's suggestion that asking questions of other men of honor is likely to be taken badly is especially interesting – it suggests that pressing other mafiosi to speak about a particular matter may often be taken as a signal of hostile intent.

[9] More generally, levels of hierarchy within Mafia families seem quite high: "Within the family, it is impossible to escape discipline or to avoid carrying out even the cruelest of orders" (Record 2451, GD, my translation).

These institutions also have a second set of problems – the enforcement mechanisms underlying the norm of truth telling are imperfect and may be nearly completely ineffective in, for example, sanctioning interfamily dishonesty or dishonesty of superiors toward inferiors. Even if the reported penalties for dishonesty are grave, there are considerable difficulties in enforcing them. Buscetta expands on his discussion of the norm of honesty to emphasize that the need for three men of honor be present for the norm to be observed stems from difficulties of enforcement. His account is worth quoting at length:

> I should make it clear that this rule obviously holds in discussions between two men of honor on their own; however, where there is no third [person] who can confirm whether a particular phrase was used or not, [the matter] rests on the word of the two men of honor; both are credible in the abstract until proven otherwise. Nonetheless, in this second hypothetical situation, it is within the power of the head of family to investigate the issue and to establish which of the two men of honor is more credible. When the discussion took place between two men of honor belonging to different families, the matter becomes more complicated, because it's only human that each head of family will tend to protect his affiliate. These questions can be brought before the *commissione*. In general, it is a good rule to refrain from talking with men of honor from families different than your own, when you are on your own; a situation might transpire where your interlocutor isn't perfectly trustworthy. (Record 1176, GD, my translation)

Similarly, one man of honor is not allowed to introduce himself as such to another man of honor, except in the presence of a third man of honor known to both, who can guarantee that the first man of honor is indeed what he says he is.[10]

These difficulties point to important weaknesses in the enforcement mechanisms underlying the norm of mutual honesty. Buscetta's testimony suggests that this norm is most likely to be enforced (1) when there is a reliable witness to what has or has not been said, and (2) where there is a credible arbiter who has appropriate incentives to sanction dishonest behavior. These two conditions are by no means necessarily easy to meet, and they are especially unlikely to apply in situations where a superior is communicating with his inferior, or where two

[10] As Gambetta (1993) notes, this norm reflects the difficulty that genuine mafiosi have in protecting their markets from imitators.

men of honor from different families are involved. As I discuss in the next section, evidence from *pentiti* and from other sources suggests that heads of family not only do not necessarily have the appropriate incentives to punish dishonesty but also may engage in direct dishonesty themselves. More importantly, as Buscetta notes, these problems are exacerbated when they involve men of honor from different families. Antonino Calderone expands Buscetta's claim substantially when he argues that mafiosi may sometimes be actively obliged to lie to other mafiosi when their immediate superior requires it:

> The bond between soldier and *decina* boss [i.e., *capodecina*] comes before everything. It comes before important obligations like telling the truth to other members of the Cosa Nostra. If a man of honor asks me about something my *decina* boss has told me to keep quiet about, I must answer that I know nothing. If the *decina* boss says to me, "Go do this murder and don't talk about it to anyone for any reason," I have to obey. I can't talk about the murder, even to the family's representative. ... And if the family boss calls me to a meeting and asks me in front of the *decina* boss whether I committed the murder, I answer, "I take his orders. Ask him about it." The *decina* boss's great power arises from the fact that the orders he gives represent the will of the family. And the family is above and beyond anything else. (Arlacchi 1993, 154)

As Calderone implies, and as Buscetta notes above, the problem of enforcement is especially likely to impinge on relations between mafiosi from different families. While interfamily disputes over who is telling the truth might perhaps be taken before the *commissione* as Buscetta implies, the *commissione* is by no means a neutral arbiter of rules, as Buscetta himself observes in a different context:

> I don't know whether these rules are always complied with ... these rules are only given lip service because in reality, the *commissione* is the instrument through which this person or that person who dominates imposes their will. (Record 1083, GD, my translation)

The *commissione*'s ability to support the enforcement of norms in interfamily matters is a function of its internal power structures and of which group or individual dominates at a particular moment in time. This has varied dramatically across different historical periods. During the late 1970s and early 1980s, the provincial *commissione* came under the influence of the *Corleonesi* and their allies, who sought carefully and systematically to undermine their rivals and expand their control to new areas. The *Corleonesi* lay the ground for their overt bid for power by systematically breaking another key norm of the Mafia – that murders should not take place on a family's territory without the permission of that family and that major murders should not happen without the explicit sign-off of the *commissione*. In a series of murders leading up to the declaration of outright hostilities, the Corleone boss Totò Riina and his allies sought to undermine other families by murdering potential rivals (sometimes on the territory of other bosses so as to foment discord) and prominent public officials without the *commissione*'s involvement.

These blatant violations of Mafia rules did not lead to any action by the *commissione*, as Riina had previously succeeded in having one of his allies, Michele "the Pope" Greco, installed as the *commissione*'s head (Paoli 2000). Greco not only ensured that no punitive action was undertaken against Riina and his allies but he also lied about these murders, one of which had involved his own nephew and subordinate, Pino "The Shoe" Greco, and had almost certainly taken place with his knowledge (Stille 1995). Stille concludes that the *Corleonesi* families' ability successfully to manipulate the rules of Cosa Nostra, and to persistently and successfully make dishonest claims without sanction was a vital element of their success:

> Part of Totò Riina's success in demolishing his rivals lay in his willingness to overturn the traditional rules of Cosa Nostra. By keeping his own membership secret [i.e., by refusing to let other families know the identities of the men of honor in his family], and lying about his actions, Riina kept everyone else off balance. (1995, 119)

In the absence of any concerted response by either the *commissione*, or by an alliance of other families, these norms were systematically traduced.

This is not the only period during which norms were set aside by actors seeking to challenge the existing dispensation; an earlier Mafia war in the 1960s witnessed similar deceptions and efforts to render an earlier version of the *commissione* redundant.[11] Whether Stille

[11] Gambetta, in an aside in the database, notes that "the mafia war in the 1960s seems to have been based on a demolition of trust within the *commissione* as a

is correct or not in his broader argument that the Mafia was permanently weakened by the *Corleonesi*'s betrayal of previously existing norms,[12] it is clear that the key mechanisms that are supposed to underpin enforcement of the norm of mutual honesty and other norms may fail during certain critical periods.

Thus, while an informal institution of truth telling and honesty exists among mafiosi (or, at least, existed during the relevant period), it did not necessarily work in a straightforward fashion. Instead, it appears to have been (1) bounded by associated norms of truncated communication, (2) superseded as appropriate by loyalty to immediate family superiors, and (3) at best imperfectly enforced in many situations.

In the final analysis, it is quite likely that the primary mechanism of enforcement for this norm was self-help. Giovanni Falcone, who professed a strong belief in this norm (which after all was a crucial precondition for much of the evidence that he had gathered to better understand the internal workings of the Mafia) perhaps puts it best when he claims that when one mafioso demonstrably lies to another, one of the two is probably not long for this world (Falcone and Padovani 1992). Here, Falcone appears to suggest that the norm is likely to be enforced not by broader social structures such as the *commissione* (there is no evidence from the testimony of *pentiti* that the *commissione* has ever actually taken action against a proven liar), but by the aggrieved party, if he is not himself killed first by the mafioso who has lied to him. If this is correct (and it seems to accord well with the available evidence), then dishonesty toward those who are about to be killed will often go unpunished for the excellent reason that the person most motivated to retaliate is now dead (although friends or relatives of those who are killed may seek vengeance for the murder itself). This provides considerable scope for mafiosi to use deception and traps to isolate and kill their enemies. I turn to this – and its consequences for trust and cooperation – in the next section.

collective organization, and among its individual members." (Record 1101, GD, my translation).

[12] This claim receives some support from the efforts of Bernardo Provenzano to reestablish traditional rules in order to strengthen the Mafia again after he assumed control.

6.3 THE CONSEQUENCES OF INFORMAL INSTITUTIONS FOR TRUST AND COOPERATION AMONG MAFIOSI

Close attention to the informal institutions governing intra-Mafia relations can help us identify one major piece of an important puzzle in the literature; the reasons for the pervasive lack of trust within the Mafia. Here, it is important to distinguish between the relations that mafiosi have with other mafiosi, and those that they have with their external clients. This last has been the topic of considerable research, in large part thanks to Gambetta's groundbreaking work on the topic. Gambetta develops an explicit model of mafioso-client relations, which demonstrates how relationships between mafiosi and their clients in broader society helps perpetuate pervasive mutual distrust and untrustworthy behavior toward those who are not protected.[13]

In contrast, the scholarly literature on the reasons for distrust *within* the Mafia is still in its very early stages of development. Mafiosi deal very differently with each other than they do with outsiders; Rocco Sciarrone (1998) argues that the mafia's relationships with external actors are usually best characterized as "weak ties" (Granovetter 1973) while its internal relationships are strong ones. Paradoxically, these internal "strong ties" go hand in hand with important forms of distrust. As Gambetta describes it, relations between mafiosi are characterized by mutual deceit and a degree of mistrust that seems often to verge upon paranoia:

> Life in the mafia is fraught with uncertainty, distrust, suspicion, paranoid anxiety, misunderstandings. Who is the strongest? Who will be the successor? Is Don Peppe really a mafioso? ... Mistakes in answering these [and other] questions occur all the time and account for many of the deaths which notoriously afflict this trade. Reading the testimony of the *pentiti*, one realizes that mafiosi scrutinize every sentence uttered by other mafiosi, searching for potential ambiguities, oblique messages, or subtle traps. (Gambetta 1993, 152–53)

Or, as the murdered magistrate, Giovanni Falcone (Falcone and Padovani 1992, 31) says, "the interpretation of signs, of gestures, of messages and silences, is one of the principal activities of the man of honor"(my translation). Although Gambetta and the sources he

[13] See Dixit (2004) for a formalization of Gambetta's arguments.

draws upon provide a rich qualitative account of the consequences of distrust within the Mafia, we still lack an account of how it intersects with the informal norms governing communication that are described in the last section.

In this section, I do not claim to provide a complete account of the sources of distrust among mafiosi, but instead to clarify the causal role that *informal institutions in particular* play in fostering distrust. As I noted at the beginning of this chapter, informal institutions are not the only important causal factor lying behind distrust. First, independent of institutions, there is a greater likelihood of clashes of interest when large sums of money and people's lives are simultaneously at stake. As the above quote from Gambetta suggests, the pervasive risk of death has a salience of its own. Thus, even apart from the relevant institutional differences, one might reasonably expect the kinds of trust and cooperation observed among, say, Bolognese machine producers, to be less likely to prevail among those engaged in criminal activities.[14]

Second, it is likely that non-goal-oriented social norms of the kind that Jon Elster (1989) describes shape trust and distrust among actors. In particular, historically rooted Sicilian norms (Pagden 1988) surrounding honor are almost certainly an important causal factor explaining the touchiness and apparently exaggerated *amour propre* of many mafiosi.[15] Furthermore, the well established norm of *vendetta* is an important vector of distrust among mafiosi; the combination of complex blood relationships and decades-long sanguinary disputes among (and sometimes within) kinship groups create an omnipresent anxiety that violence may be renewed at any point. These norms may shade at the edges into more rationally grounded informal institutions

[14] This is not to say that they are impossible; in a fascinating article, Federico Varese (2006) discusses how operators in the heroin market in Verona "belonged to the same social milieu that had given rise to a flourishing economy and adopted the same entrepreneurial spirit and straightforward commercial practices that characterized the legal sectors of the economy. Transactions in the illicit drug market took place according to shared rules of fair bargaining, and punishment took the form of exclusion from future exchanges and refusal to offer credit and discounts. In addition, a significant level of barter and individualized exchange existed in this market, and no third-party mechanism to punish defectors existed" (Varese 2006, 435).

[15] Pagden (1988) suggests that these norms were deliberately instilled under Hapsburg rule in order to prevent threats to the dominant social order from arising.

(as Elster's discussion implies, nonconsequentialist norms and informal rules based on self interest and the fear of punishment are often intertwined with each other), and may also sometimes serve as justification for materially motivated actions. They do nonetheless have an important independent causal role in shaping actors' behavior.

Thus, my goal in this chapter is not to provide a complete account of trust and cooperation (or the lack of same) between mafiosi. Instead it is to assess whether and how the informal rules governing communication that I have discussed in the previous chapter impact trust and trustworthiness among mafiosi. More specifically, my arguments would suggest that informal rules of honesty among mafiosi will affect actors' propensity to trust or distrust others, to the extent that they provide actors with grounds for assessing other actors' motivations. As I argue below, the informal institutions shaping information exchange among mafiosi are likely to promote distrust by actively helping further paranoia and confusion about people's underlying motives.

Why do the relevant informal institutions have these consequences? On the face of it, they do not seem entirely dissimilar to the informal institutions mandating honest dealings, which, I have argued, support quite widespread forms of trust and cooperation among manufacturers in Bologna. My argument stresses two elements. First, the informal rule of truth telling among mafiosi goes hand in hand with ancillary norms that tend to prevent straightforward communication. Far from casting a benign penumbra that might help ground the beliefs of mafiosi over the trustworthiness of others, these secondary norms are likely to cloud and obfuscate those actors' actual intentions and motivations, making it difficult for others to trust them.[16] Second, weaknesses in enforcement, combined with both the malleability of these forms of communication themselves and the active expectation that mafiosi should lie to each other under some circumstances, not only are likely in themselves to lead actors to distrust each other rather than to trust each other in many circumstances, but also allow some mafiosi directly to manipulate others' expectations so as to increase distrust among them directly.

[16] Clearly, when I use the term "benign" here, I am only referring to its consequences for trust, not the socially malign forms of cooperation that this trust might support.

As Chapter 2 discusses, institutions are likely to affect trust between actors to the extent that they provide actors with grounded reasons to believe that others will behave in a trustworthy fashion toward them. The institutions governing communication among mafiosi do allow for the dissemination of certain kinds of information, albeit with frequent elisions and distortions. Obvious examples include information on who the members of the mafia are, which families control specific neighborhoods, and, on a need-to-know basis, who has been responsible for which actions.

However, these institutions are likely to be *especially* bad at transmitting the kinds of information about actors' underlying motivations that are most crucial to creating and sustaining trust. The combination of deliberate terseness, manipulation of communicative ambiguities and threatened punishment for those who seek to push certain lines of inquiry too far leads to a kind of communicative equilibrium in which the motivations and intentions underlying actions will rarely be clarified by the authors of those actions (although third parties may surely try to infer them). This is one of the main reasons why the mafia is, as Rocco Sciarrone (1998, 41) describes it, a low information environment characterized by "a strong ambiguity over preferences."

Even in the best of situations, mafiosi are likely to be highly uncertain regarding the intentions and motivations of other mafiosi. For example, Antonino Calderone (Arlacchi 1993, 263) describes how difficult it was to discern the reason why Riina, who had recently ordered the murder of Calderone's brother, delivered a glowing encomium to the dead man at a meeting of the *commissione*:

> It was difficult to tell whether ... his elevated and noble words were coming from sincere grief ... or from the base satisfaction of a victor who has just eliminated a dangerous enemy ... It was useless to try to dissect, understand, make sense of it. It's always like that in the Cosa Nostra: no fact ever has only one meaning.

In situations where there is some reason to suspect that others may have a grounded reason for behaving in an untrustworthy fashion, this uncertainty is likely to tip over into active distrust. Just such a reason is provided by the second factor that my explanation stresses; the combination of (1) systematic limits to the enforcement of the informal rules governing truth-telling, and (2) the considerable scope for dissimulation that is provided both by ambiguous communication and direct lying. Together, these create a communicative environment, which is fertile ground for deceit and treachery.

The confessions of *pentiti* and other sources of information provide ample evidence that deception is a preferred means to isolate your opponent and murder him. As Giovanni Falcone, the murdered anti-Mafia magistrate notes, mafiosi have little interest in dying in pursuit of their objectives. Where it is necessary or convenient to kill other mafiosi, they prefer to use subterfuge to entice their victims to places where it is easy to kill them with little fuss, ideally through means such as strangulation that are silent and leave little forensic evidence (Falcone and Padovani 1992).

Thus, for example, the mafia boss Rosario Riccobono was rewarded for betraying his close friend Emanuele D'Agostino to the *Corleonesi* by being invited to a barbecue at Michele Greco's estate, where he himself was awakened from his habitual postprandial nap by a group of killers with the strangling cord in their hands (Stille 1995). Giovanni Brusca (Lodato 1999) says that he has it on good authority that Stefano Bontate was murdered by people he knew well, and trusted enough to invite into his house and make coffee for, before they killed him. The *Corleonesi* seem to have been especially adept at this kind of betrayal, but other mafiosi also frequently use ties of friendship to lure victims to places where they can easily be killed (Falcone and Padovani 1992). For example, when Mario Di Bella, better known as "The Pig," got into trouble with another mafia family, he was tricked into going to a place where he could be killed:

> At this point Giuseppe Madonia asked the Catania family to give him Di Bella. As usual, deception was employed. Salvatore Marchese knew Di Bella well, and could employ himself as bait. ... [At a barber's shop], he suggested to Di Bella that he had acquired some stolen paintings, and they decided to go to see them. Needless to say, poor Di Bella got bullets, not paintings. (Record 1746, GD, my translation)

Similarly, Nino Matranga, an "old-time provincial representative who was held responsible for the first mafia war," was killed by Stefano Bontate and others, who used Matranga's brother-in-law in whom "he placed a lot of trust" to lure him into a trap in Milan (Record

1670, GD, my translation). Many other stories from the testimonies of *pentiti* illustrate this general theme.

Deception of this kind creates an environment that is highly conducive toward distrust among mafiosi. If mafiosi (1) have few reliable means to discern the actual motivations of other mafiosi, (2) are aware that other mafiosi frequently engage in deceptive communication so as to further their own interests without being effectively sanctioned under existing informal institutions, and (3) know that such deceptions often lead to the death of incautious and unsuspicious mafiosi, then we may reasonably expect mafiosi to distrust each other over a wide variety of issues. The relevant informal institutions provide mafiosi with little reason to trust other mafiosi as a general class of actors over a broad set of issues, and strong and cogent reasons to distrust them. Moreover, not only do these institutions provide little information that would lead mafiosi to trust in the motivations of specific others but they may also actively undermine such trust by instantiating a set of expectations under which even apparently trustworthy individuals – friends, close co-workers, and relations by blood or marriage – may lie in order to trap the mafioso and kill him. As Giovanni Falcone describes the world of the Mafia:

> Inside an organization like the Mafia, one assesses things in a different way than outside. Something about a violent death that might horrify people either as magistrates or as ordinary citizens – a man meeting his death at the hands of his best friend, the strangling of one brother at the hands of another – doesn't produce the same reaction among men of honor. ... For men of honor, what is important is the courage and professionalism demonstrated in the killing. ... Cosa Nostra is founded on the rule of obedience. He who knows how to obey, carrying out orders with the greatest expedience possible, has assured his career. (Falcone and Padovani 1992, 31, my translation)

Falcone argues at length that Mafia rules make it nearly impossible for mafiosi to evade orders from their family superiors to kill others, even if these others are close friends or relatives. The constant fear that one's friends or family may prove untrustworthy conducts toward the kind of diffuse distrust and extreme wariness of others' motivations that Gambetta describes. Pervasive distrust may perhaps tip over into the kinds of paranoid cognition described by Roderick Kramer (1994), but it is in general a plausible and entirely rational response to the social environment that mafiosi inhabit. Overly trusting mafiosi are likely to end up dead. As an anonymous *pentito*'s account makes clear, even wary mafiosi may be killed if they do not anticipate betrayal at a particular conjuncture:

> One day, the *capodecina* told the anonymous informant and another member to carry out a mission. This involved eliminating someone who was hiding in a farmhouse ... on the morning of the day for the ambush, the *capodecina* sought out the informant to clarify the mission for him. There wasn't anyone hiding in the farmhouse, and the objective was to eliminate the other soldier, who suspected nothing. As it transpired, the two arrived at the farmhouse, and the informant stayed back to "cover their backs" according to the official story; the other man had hardly entered the farmhouse before he was shot with a pistol. The informant never asked, and never knew what the motive for this action was ... "Tano [the other soldier] however had never even understood what was happening: this was the one thing that was certain, and was the thing that gave me sleepless nights. Tano was crafty but he had never understood that his 'party' was being prepared. I however would have done the same thing, in his place. Now I knew that it could happen to me."[17]

Indeed, senior mafiosi seem to have few compunctions about murdering their underlings on the basis even of vague suspicions as to their ultimate intentions:

> Stefano Giaconia was one of the most notable men in the Palermo-Centro family during La Barbera's reign. When he became part of the Bontate family, he didn't, however, set aside the ambition to become the head of the Palermo-Centro family some day when it had been reconstituted. One day, Stefano Bontate was late for a meeting at Michele Greco's house. When he arrived, he apologized, and said that he had been inconvenienced because he had had to change his tire. He added that he had strangled Stefano Giaconia because the latter had made him feel worried. Michele Greco replied that he had done the right thing. (Record 1686, GD, my translation; see Arlacchi [1993] for further detail)

[17] See also Giovanni Falcone's (Falcone and Padovani 1992, 35, my translation) account of how the Mafia *commissione* entrapped Michele Cavataio: "a certain number of men of honor were instructed to make themselves friends of this terrifying and extremely untrusting person ... This decision wasn't taken lightly, because Cavataio was widely admired for his courage. But Stefano Bontate, head of the Santa Maria di Gesu family had the better of the argument ... This two-pronged strategy – friendship and elimination – was strikingly successful: Michele Cavataio was brutally murdered in December 1969, in an ambush that has passed into history as the battle of viale Lazio."

It is at the very least highly plausible that the demonstrated willingness of bosses to eliminate potentially "worrying" underlings without further explanation generates suspicion, wariness, and distrust among those underlings.[18]

As discussed in the previous section, existing norms about the dissemination of information create gaps in communication that Mafia superiors are in an especially strong position to exploit. While superiors receive information both from those beneath them and from contacts with other families, they may disseminate this information as they see fit. This means that they may also selectively share information or disseminate disinformation so as to divide their underlings and make them more likely to distrust each other. For example, Paoli (2000) discusses the ways in which Mafia bosses seek to foment distrust among their underlings in order to prevent them from forming horizontal alliances:

> As Vincenzo Calcara reports, every now and then Capogreco, his godfather, told him that another "man of honor" had complained about him, asked him to follow the movements of another associate, or made him suspect that others were checking on him, too: "He stirred up trouble and invented 'tragedies,' so that no positive feeling would come into being between us soldiers. The family and the rules had to come first. ... I always had to be on the alert, to distrust everybody and answer only to Capogreco." (Paola 2003, 93)

These efforts to stir up mutual suspicion among subordinates may even lead to murder, as Antonino Calderone suggests in his (perhaps not entirely reliable) version of a famous killing that smoothed the ascent of Luciano Leggio and his lieutenants Totò Riina and Bernardo Provenzano to power among the *Corleonesi*, that the killing was in part provoked by a certain "Funcidda," who "spread false rumors about the two [Leggio and the murdered mafioso] in order to make them hostile to each other and stay in power" (Record 1911, GD, my translation).

Similarly, mafiosi in privileged positions may try to exploit gaps in intrafamily communication so as to exacerbate tensions among them. According to Tommaso Buscetta, the proximate origins of the first Mafia war lay in efforts by malefactors to stir up distrust

[18] Underlings too may of course seek to betray and eliminate their bosses; see the final chapters of Arlacchi (1993).

between various families, and in particular to weaken Angelo and Salvatore La Barbera, who were becoming too powerful.[19] On the one hand, Michele Cavataio had an enemy of the La Barbera brothers murdered, so that the Barberas would be suspected of having ordered the killing, leading to vicious internecine battles (Stille 1995). On the other hand, Pietro Torretta sought to increase the tensions further by persuading the brothers that the rest of the Mafia had turned against them:

> Pietro Torretta, head of the Uditore family, and member of Salvatore Manno's district, developed a plan to turn Angelo La Barbera's anger against the *commissione*. He lured two of La Barbera's men to his house with the intention of killing them in secret and making them disappear. La Barbera would certainly have blamed the *commissione* for this "if it hadn't happened that the two young men sought to flee, and one of them jumped from the balcony of Torretta's house, therefore making it necessary to eliminate the two thugs in a blatantly obvious hail of gunfire." (Record 1101, GD, my translation)

These efforts sparked a vicious internecine war, which Cavataio sought to keep stoked up by periodically assassinating figures on both sides of the dispute. This war certainly was not a simple product of deficiencies in the informal institutions governing communication among mafiosi; its underlying cause was the division of the spoils from drugs money. However, it was exacerbated by the limits of these norms – mutually reinforcing suspicions and paranoid expectations between different Mafia families were reinforced by the inability of families credibly to communicate over what they had or had not done, and who they had or had not killed, helping to fuel a self-reinforcing cycle of violence.

Thus, informal institutions are a key vector and cause of distrust among mafiosi. First, they lead to truncated and ambiguous forms of communication, which make it very difficult indeed to discern actors' true intentions. In a world where there is ample evidence that other mafiosi will often have malign intentions, and where actions taken on foot of these intentions often have unpleasant or fatal consequences for the unwary mafioso, mafiosi will unsurprisingly tend toward suspicion and distrust, rather than trust, across a wide variety

[19] However, one should note that it is not clear that Buscetta's account of the first Mafia war is entirely honest. John Dickie (2004) suggests that Buscetta played a more ambiguous role than he implied in his testimony.

of interactions with other mafiosi. Second, apparent exceptions to the norm of truth telling permit the use of deception in order to trap and kill other mafiosi. Such deception appears to be very nearly systematized – Calderone, Falcone, and others suggest that deceit is a tactic of first resort when mafiosi are confronted with the need to kill another mafioso. There is no evidence from the records of it being punished by anyone other than the intended target (should he survive), and much evidence to suggest that it is regarded as acceptable, and perhaps even appropriate by uninvolved third parties in the Mafia under many circumstances. The likelihood of such deception (and the apparent rarity of punishment) provides good reason for mafiosi to distrust other mafiosi's promises and commitments.

Finally there are asymmetries, gaps, and imperfections in the norms governing communication (superiors may choose whether or not to disclose information to inferiors while inferiors often have little leverage over superiors and their associated sanctioning mechanisms; communications between families are likely to be sparser and more subject to distrust than communications within them), and in the sanctioning mechanisms associated with them (superiors are unlikely to be punished for misleading inferiors; dishonest communications between mafiosi from different families are difficult to punish). These asymmetries, gaps, and imperfections provide ample room for privileged mafiosi to manipulate information flows so as directly to increase distrust among other mafiosi, hence creating divisions that can be exploited for gain. None of this is to argue that distrust among mafiosi can be *reduced* to the effects of informal institutions governing communication and other forms of behavior. Other causal factors (the risk of death, the sensitivity of the matters being transacted over, the existence of non-goal-oriented norms) are important, both in terms of their effects on the relevant institutions and in their own right. Even so, the relevant informal institutions – and their limitations – have important causal consequences.

6.4 CONCLUSIONS

Informal institutions may not only coexist with active and pervasive distrust among individuals over a broad variety of issues but may also indeed help cause it. Informal institutions may exacerbate ambiguities over actors' ultimate intentions. They may permit (and indeed implicitly reward) deception and treachery under some circumstances. Finally, they may create gaps in communication that allow privileged actors to manipulate other actors' mutual expectations.

As I have argued throughout, it is not possible to conduct a proper controlled comparison of the informal institutions that lead to pervasive distrust among mafiosi over many important issues with the informal institutions that support trust and cooperation among Bolognese business actors. In other words, one cannot reasonably attribute differences in the broad outcomes of the two cases (differences in patterns of trust and distrust) to differences in the explanatory mechanisms (informal institutions), given that there is uncontrolled variation in a wide variety of other factors that are arguably causally important. The most one can do is (as I do in this chapter) to use process tracing to substantiate the claim that the particular institutional mechanisms in question appear, *ceteris paribus,* to increase or decrease trust in important ways, without claiming that the gross differences in outcome between the cases are entirely attributable to these mechanisms. Nonetheless, it is both possible and helpful to examine similarities and differences in how the relevant causal mechanisms work in the two cases so as to help generate theoretical insights that may be tested in future work.

In the Bolognese packaging machinery district, the relevant informal institutions appear to have worked in two ways to support trust over important aspects of subcontracting relationships. First, these institutions instantiated a general set of expectations over how final firms (with the exceptions of the very largest firms in the district) were supposed to behave toward subcontractors, and vice versa. Under these institutions, final firms (as a class) had a good idea in general terms of what they could and could not expect from subcontractors (as a class) in imperfectly anticipated future situations, and subcontractors similarly had a good idea of what they could expect from final firms. Because these mutual expectations conducted toward trustworthy relations (and were accompanied by the credible threat of retaliation in cases where either party misbehaved), actors belonging to each class had good grounded reason to trust actors belonging to the other class in general.

Second, the relevant institutions also provided means of circulating information about the behavior of *specific* actors within the community. Actors who deviated from the relevant community norms were quickly and (relatively easily) identified through gossip, and they acquired a bad reputation, which hurt their chances of doing business in future.

Both these institutional mechanisms conducted toward trustworthy behavior and trust across a wide variety of fuzzily defined situations. The relevant institutions, then, both embodied information (generalized expectations about how actors of specific classes were supposed to behave) and transmitted it (specific information over the behavior of particular actors) in a relatively transparent and useful way.

Among mafiosi, in contrast, the relevant informal institutions worked so as to make trust among mafiosi less likely and distrust more likely than would otherwise have been the case. The relevant informal institutions provide little reason for mafiosi to trust each other as a class over many important intra-Mafia issues, and much reason to distrust each other. Even though an informal rule of truth telling appears to apply part of the time, it is at best imperfectly enforced and appears to have imposed few constraints, if any, on mafiosi seeking to use deceitful communication and treachery to entrap and kill each other when it appeared advantageous. Furthermore, mafiosi are expected to kill others – even close friends or relations – should the family so require. The result is pervasive suspicion, mutual monitoring, and profound distrust that sometimes verges upon paranoia.

Similarly, the relevant informal institutions appear to have been ill suited to provide information on the motivations and intentions of particular mafiosi, let alone to anchoring for trusting relations. Informal norms mean that communication among mafiosi is often ambiguous and elliptical and strongly discourage the answering of questions. This makes it harder for mafiosi to discern each others' intentions, and thus harder for them to trust each other in the grounded ways that Hardin discusses. More generally, as discussed above, the inadequacies of enforcement mechanisms and gaps in the communication structure both mean that it is difficult for mafiosi to trust in each other's honesty, and they provide ample scope for privileged actors to strategically manipulate information flows so as to increase distrust among others in ways that redound to their own advantage.

In very broad terms, informal institutions in both cases act according to a similar logic. That is, they provide information to actors about the likely behavior of others both as members of broader social classes or categories and as specific individuals with particular intentions and motivations. However, their apparent consequences are quite different. Informal institutions of the kind observed in Bologna cast a shadow over a variety of future situations that makes actors far more likely than they would be otherwise to trust, and hence to cooperate, with each other. Informal institutions of the sort that prevail among mafiosi similarly cast a shadow over a variety of future situations, but with the opposite effect. They perpetuate distrust and make actors less likely to cooperate than they otherwise would be.

This suggests that the theoretical arguments set out in Chapter 2 are likely to travel well to contexts other than business relations in industrial districts. The mechanisms producing distrust among mafiosi and trust among Bolognese machinery manufacturers have very different consequences. Nonetheless, they display a strong familial resemblance to each other. By paying close attention to the ways in which informal institutions (1) provide general information about the motivations and behavior of broad classes of actors and (2) undergird systems of communication that disseminate particular information about the past behavior and likely intentions of specific others, we can arrive at useful conclusions about how these institutions can produce not only trust but distrust. The arguments set out in this chapter are only a first step. More controlled comparisons are necessary to map out the scope conditions under which relevant informal institutions are likely to lead to the one or the other outcome. Even so, the clear relevance of these arguments to a very different set of relationships among actors in a quite different setting both provides some degree of confidence that the theory has more general application, and shows how it may help explain not only trusting relations but the generation of distrust too.

7

Conclusions

7.1 INTRODUCTION

In Chapter 1, I argued that trust and cooperation are central to our understanding as political scientists of how political and economic systems work. Nonetheless, the sources of trust and cooperation are very poorly understood. The study of institutions is now at the core of our discipline, yet we know little to nothing about the relationship between institutions and the richer forms of shared expectations that we commonly group together under the rubric of trust. When do institutions produce, or support, trust? When do they produce or perpetuate distrust? Or do institutions, trust, and distrust have any relationship at all?

We do not have good answers to these questions, in large part because of underlying theoretical confusion as to what trust and cooperation (and, indeed, institutions) consist of. Our major explanations of trust and cooperation tend variously to leave institutions out, to reduce trust to a mechanistic product of institutions, or to claim that there is a two-way relationship between institutions and trust without really explaining how this relationship works. As a result, while there is a substantial body of empirical literature which claims that institutions affect trust, we know next to nothing about the specific mechanisms through which one or another set of institutions might have one or another consequence for the ways in which individuals trust each other.

In this book, I have sought to provide an account that will bridge the gap between institutions and trust by showing how institutions may provide a basis for quite complex forms of trust and cooperation. I have also sought to show how an account of this kind can help explain forms of trust that have previously eluded rational choice theory, by taking on the "hard case" of inter-firm cooperation in industrial districts. A demonstration of how institutions affect trust in these districts illustrates the value of combining an account of trust based on individual interest (Hardin 2002) with a theory of how institutional rules shape behavior when actors are dealing with others whose specifically personal qualities are not necessarily well known, but who fall into "types" described under the relevant institutions. Finally, I have shown how institutions can help perpetuate expectations that foster distrust too.

In order to develop this theory and its empirical consequences fully, I have sought to build a cumulative account, using the arguments laid out in Chapter 2 as the foundations for a series of successively more complex arguments about how trust works. Thus, the book seeks to speak to three overlapping debates – debates on the sources of trust, comparative debates on how trust works in different settings, and debates on the relationship between institutional change and trust. In the remainder of these conclusions, I lay out in turn the contribution that my arguments make to each of these debates.

7.2 THE SOURCES AND NATURE OF TRUST

I have argued for a version of Russell Hardin's (2002) "encapsulated interest" account of trust. As I hope I have demonstrated, this account provides a more supple and useful account of trust than do rival accounts that deny or minimize the role of self-interest in trust relations.

First, it allows us to describe trust relationships precisely. Rather than relying on generic notions of "diffuse interpersonal trust," which are arguably at least as much an artefact of survey questionnaires as a measure of underlying relationships, it relates trust relentlessly to the particular. In this, it is more faithful to our everyday intuitions about what trust involves – few, if any of us accord unreserved trust to others in our society with regard to every matter, no matter what we may say to researchers with badly phrased questionnaires. When we trust or do not trust others, we trust or do not trust them with regard to particular matters (or, in my extended arguments, matters of a certain kind). Hardin's language, which allows us to distinguish between different potential trusters, different potential trustees, and different matters over which these trusters do or do not trust potential trustees, provides a major advance in our understanding of how trust operates.

Second, and more pragmatically, it provides an account that accords well with individuals' own descriptions of how they do or do not trust others, as demonstrated by interviews with business actors in both case studies. Interviewees referred again and again to others' interests as providing reason for trusting or distrusting them. They tended to trust those who had an interest in behaving in a trustworthy manner, and to distrust those who did not. This is not necessarily a guarantee that the encapsulated interest argument actually describes these individuals' underlying thought processes: individuals may not themselves understand the mechanisms through which they reach decisions. Still, at the very least, claims consonant with individuals' self-understandings face a much lower burden of justification than do arguments at odds with these self-understandings. The latter not only have to explain why individuals actually take the decisions that they do but also why they are mistaken about the forces driving these decisions.

In short, the encapsulated interest account of trust is not only of interest to political theorists. It is an extraordinarily valuable tool for comparativist political scientists and, indeed, for all political scientists who might wish for an account of trust and cooperation with strong microfoundations. The case studies examined in this book should be hard cases for the encapsulated interest account. They involve varieties of trust and cooperation that have been repeatedly described as nonrational by the scholars who study them. Yet as this book demonstrates, the encapsulated interest account, when combined with institutions, provides an excellent basis for understanding the circumstances under which actors in these industrial districts do or do not trust and cooperate with each other.

This said, the somewhat simplified form of the encapsulated interest account that I present here does not provide a complete explanation of trust and cooperation in the two case studies. In order to provide simple and clear predictions, I combine the encapsulated interest account with a relatively stringent set of assumptions about rationality. I assume that individuals are motivated by material self-interest: they seek to maximize their wealth. This only provides a simplified account of individuals' motivation and, in some circumstances, will surely make incorrect predictions.

The kinds of arguments that I offer could be fruitfully complicated in at least two ways. First, one might introduce motivations other than narrow material interest as part of one's explanation of why actors behave the way that they do. Thus, one could reasonably surmise that individual artisanal producers are driven to set up their own firms not only by the desire to make more money but also by the desire to be more independent – to be their own bosses (Bertini 1998). If this is true, they may be less likely to cooperate in ways that circumscribe their autonomy as producers. One might also predict that individuals will under some circumstances cooperate with others, not because this is economically advantageous, but because the others are friends whose friendship is valuable on its own terms. These motivations surely exist and might help provide the basis for a fuller account of the sources of trust than the account I present here.

As an alternative to examining different motivations for individual behavior, one might seek a more accurate understanding of the cognitive processes underlying trust by identifying and elaborating the relevant psychological mechanisms. There is a rich literature in social psychology that does exactly this.[1] This would provide a better understanding of how individuals' psychological states may affect their trust or distrust of others, as well as provide a model of cognition that is truer to life than the somewhat implausible assumptions of game theory. For example, Roderick M. Kramer's work on "paranoid cognition" (1994; 2002) shows how individuals may not be able to interpret others' intentions correctly if they start from the premise that others are probably out to get them. Here, the ability of actors to make rational judgments on the basis of observable facts about others' behavior may be distorted in important ways.

Either of these modifications would make it easier to incorporate forms of trust and cooperation (and distrust and failure to cooperate)

[1] For useful surveys of this literature, see Kramer 1999 and 2002.

that are not well explained by the framework offered in this book. Equally, they would make it more difficult to make clear predictions about when individuals will or will not trust each other. Given these tradeoffs, a simpler framework, such as the one offered in this book seems more closely suited to the purposes of predictive social science.[2] It allows one to make clear predictions about how actors are likely to behave given observable conditions, even if these predictions are likely sometimes to be falsified. As long as the mechanisms emphasized in this framework do play an important role in individuals' trust or distrust of others, it will still have substantial explanatory value.

In short, the empirical evidence set out in this book shows how the encapsulated interest account of trust provides a good description of how people actually think about trust and distrust in the political economy and the basis for broader predictions about when they will trust or distrust each other. As such, it provides an excellent foundation for arguments about trust and cooperation. Given the rather scanty microfoundations of dominant alternative approaches, such as the voluminous political culture literature on trust, this is an important theoretical advance.

7.3 INSTITUTIONS AND TRUST

The arguments advanced in this book differ from those of Hardin and his colleagues in one important way. Rather than seeing institution-supported expectations and trust as substitutes for each other, it argues that institutional rules can be a key source of trusting relations among individuals. In an important recent critical overview of the literature on trust and cooperation, Karen S. Cook, Russell Hardin, and Margaret Levi (2005) make a strong argument for a theory of trust that is grounded solely in relationships between individuals. They claim that

[2] Scholars who prefer more inductive or descriptive forms of social science may very reasonably disagree, preferring a richer and more detailed account over a sparser "cleaner" one. In making the argument that I make here, I do not especially wish to engage in the metadebate over which is "better" or "worse." Instead, I am making the more limited claim that a radically simplified account of human interaction is appropriate and reasonable for scholars wishing to engage in certain forms of social scientific inquiry, even if it necessarily leaves out some of the complexities of real world human behavior.

> *trust involves a genuine involvement between you and the trusted other and a specific, not abstract, assessment of that other's motivations toward you.* If you have never met or in any way dealt with the other person ... it is meaningless to say you trust that person. The word *trust* has no real meaning if it does not differentiate our relations with others because trust is inherently about our relationships. (Cook, Hardin, and Levi 2005, 8, italics in original).

This means that, in their view,

> trust plays a relatively small role on the grand scale in producing and maintaining social order. We usually rely on and cooperate with each other, not because we have come to trust each other, but because of the incentives in place that make cooperation safe and productive for us. We can take risks on one another even as complete strangers in many situations because of the presence of institutions, organizations, and networks that constrain exploitation and provide individual and collective incentives that reduce externalities. (Cook, Hardin, and Levi 2005, 14–15)

Under this set of arguments, trust is an insufficient basis for widespread cooperation in advanced industrialized societies. Instead, more impersonal structural supports for cooperation are appropriate. Under Cook, Hardin, and Levi's account, the two may surely interact: institutions such as laws may make trust easier by providing a level of background enforcement that makes it easier for individuals to trust each other over smaller issues in their personal interactions. Nonetheless, the two support cooperation in very different ways.

This perspective clearly has very substantial merit. It distinguishes trust from mere institutional compliance and highlights the many ways in which institutions may produce social cooperation without directly producing trust. However, it faces some difficulties in accommodating informal institutions, which play a pervasive role not only in less developed countries but also in advanced industrial democracies. Should the kinds of cooperation produced by these institutions – which do not involve the same, quite narrowly based expectations as formal institutions – be characterized as trust?

For their part, Cook, Hardin, and Levi suggest that trust is unlikely to be implicated in the kinds of small, densely interlinked communities where we might expect informal institutions to play an important role. Indeed, they de-emphasize the role of informal institutions in these communities, arguing that action is likely to be motivated by nonrational norms (Elster 1989) rather than by rational consideration of what others' interests involve.

However, as I argue in Chapter 5, this does not seem to describe very well how actors behave in the Bolognese packaging machinery district, which arguably constitutes just such a densely interlinked community of actors. Informal rules play an important role in this district, as they do in other Italian industrial districts, but they do so because of the ways in which they affect actors' interests and their perceptions of those interests. Packaging machinery manufacturers seem quite aware of what their interests and the interests of other relevant actors are, and they structure their relations accordingly. Thus, while informal rules play a highly important role in these districts, they serve less as nonrational motivations for action than as rules that structure the rational beliefs of actors regarding their own and others' interests.

This is certainly not to exclude the possibility that norms, in the strong sense of the term that Cook, Hardin, and Levi employ, may play an important role in many communities. But it is to argue that there is an important set of interactions between informal institutions and trust that cannot be very well captured by a purely relational model. Indeed, as Cook, Hardin, and Levi acknowledge elsewhere in the book, informal cooperation in "high-trust" organizations relies on a mixture of iterated relationships over time and "self-enforcing rules and norms" (Cook, Hardin, and Levi 2005, 145). The two may reinforce each other: trusting relations in a workplace are not simply a congeries of discrete relationships between and among individual workers and individual supervisors. Instead, they are deeply embedded in shared, informal expectations (Miller 1992) about what workers (as a class of individuals) and supervisors (as a class of individuals) should do across various broadly defined situations. These expectations in turn are instantiated in informal rules. Clearly, then, there is a class of trusting relations that involve something very similar to the encapsulated interest account of trust, but that are difficult to capture fully within Cook, Hardin, and Levi's framework. On the one hand, they rest to a very large extent on institutionalized rules, on the other hand, they involve a much greater degree of trust-like expectations than we usually associate with institutions.

This ambiguity can be addressed if we acknowledge (1) that institutions *can* directly impinge on trust relations but (2) that certain institutions (informal institutions) are likely to have much more important direct consequences for trust than others. As I have argued, we can accommodate the effects of institutions on trust by extending Hardin's encapsulated interest account, so that it not only involves personal relations but also broader expectations about the ways in which different classes of actors are likely to interact in specific social situations. As well as enquiring into whether actor *a* trusts actor *b* with regard to matter *c* because the relationship between *a* and *b* gives *a* grounded expectations that *b* will be trustworthy, we may ask whether actors of class *A* may have grounded expectations that actors of class *B* will be trustworthy in situations of type *C*. Expectations at this level of generality will not be grounded in specific personal relationships but in institutional rules that provide information about how actors of class *B* are likely to behave toward actors of class *A* in situations of type *C*, or in rules over communication that allow particular information about particular others to be conveyed. Where these rules are narrow and clearly defined, trust will not be implicated. Instead, to the extent that these rules conduct toward cooperative behavior, this may happily be accommodated as one variety of the institution induced cooperation that Cook, Hardin, and Levi rightly distinguish from trust-based cooperation. When, in contrast, the institutions are relatively loosely defined beforehand, as with the informal rules and norms that govern cooperation in many densely structured communities, we may very reasonably describe the outcome as institution-induced trust. Institutions of this sort produce a set of expectations that are recognizable as trust – they have that degree of imprecision that Gambetta and others claim is important to trusting relations. However, they do not rely on purely personal relationships to produce that effect. Thus, I suggest that the account I lay out may very usefully supplement and modify the encapsulated interest account. On the one hand, it maintains the distinction between trust and broadly impersonal forms of cooperation that Cook, Hardin, and Levi are at pains to stress. On the other hand, it better accommodates important forms of trust that *do* seem to rely directly to a greater or lesser degree on rules as well as (or even instead of) personal relationships.

Furthermore, because the framework laid out in this book allows one to make positive predictions about the relationship between institutions and trust, it provides the basis for a *comparative* account of this relationship. Specifically, as discussed in Chapter 4, it predicts that differences in institutions will be associated with differences in the ways that individuals trust or do not trust each other. Political systems at the national, subnational, or community level, in which effective formal institutions predominate, will ipso facto not be systems where individuals need to rely on trust relations in order to cooperate. In contrast, political systems without effective formal political institutions *will* require individuals to rely on trust (whether this trust is underpinned by personal relations, informal institutions, or a combination of the two) if they wish to cooperate over relevant matters.

Certain informal institutions *are* likely to support a high level of trust over important matters among actors in the relevant community. This is clearly the case in the Italian case study, where there were widespread forms of productive cooperation outside hierarchy that appeared difficult to achieve in the German case. More generally, we may say that for certain kinds of cooperation, appropriate informal rules provide a degree of flexibility that would be impossible to achieve using formal mechanisms. This echoes and extends Gary Miller's (1992) more general claim that formalized Taylorist incentive schemes are incapable by their very nature of achieving certain efficiencies. The kinds of high-cooperation, high-trust environments that Miller identifies may be supported by appropriate informal rules that provide the relevant actors with a high degree of security that their cooperation will be reciprocated even in unexpected circumstances.

Nonetheless, informal institutions by no means necessarily support high levels of trust. Where informal institutions that might underpin trust and cooperation do not exist, we may reasonably expect that trust will be confined to close familial or other relationships. Furthermore, some kinds of informal institutions may actively conduct away from trust and trustworthiness. Here, for example, we see how the rules governing relations among mafiosi in Sicily conduct away from trust. They support generalized expectations over the ways in which mafiosi behave toward each other in a wide variety of situations, which strongly suggest that other mafiosi cannot be trusted. And they lead to truncated forms of communication, in which actors' real motivations and intentions are obscured, and which furthermore offer extensive opportunities for manipulation of information flows so as to create distrust among others.

The account developed in the course of this book shows how institutions may have important consequences for trust among individuals, and thus for cooperation. Not only may institutions underpin "cooperation without trust" as Cook, Hardin, and Levi describe it but they may also serve as the basis for important forms of trusting cooperation too. This provides the necessary microfoundations for comparisons between different institutional systems and their effects on trust. As Chapter 4 shows, we may expect important differences between institutional systems where formal institutions predominate and those where informal institutions play a key role. And as Chapter 6 demonstrates, we may also expect that there will be differences between different kinds of informal institutions in terms of their effects on trust.

7.4 INSTITUTIONAL CHANGE, TRUST, AND ECONOMIC COOPERATION

Finally, the arguments set out in this book have implications for current debates about institutional change, trust, and cooperation. It sets out a framework for understanding how changes in the bargaining power of actors may have consequences for institutional rules, and how these rule changes may in turn have consequences for the ways in which individuals trust and cooperate with each other.

Specifically, the framework presented in Chapter 5 provides an alternative to current understandings of institutional change and trust, as well as of institutional change in the political economy more generally. The literature on how institutional change may impinge on trust relations is sparse. Much of it relies explicitly (Franchi and Rieser 1991) or implicitly (Harrison 1994) on very general arguments about social change, most notably Tönnies's (1887) distinction between *Gemeinschaft* and *Gesellschaft*. Like many other major contributions to sociological theory from the nineteenth century, this dichotomy reflects a historically specific set of concerns about the social disruptions incumbent on industrialization and population movements from the country to the city (Tilly 1989). It suggests that there is a sharp distinction between two fundamentally different forms of organization and that a irreversible historical shift has and is taking place from the one to the other. It thus has a tendency to overstate the profundity of the differences between cooperation based on informal institutions and cooperation based on formal ones; a careful examination of the historical record (Greif 2005) suggests that while there are important differences between the two, they are both based on rules and enforcement mechanisms. Finally, arguments of this kind imply (often incorrectly) that forms of trust and cooperation based on informal institutions are likely to wither away as modernization proceeds.

But, there are accounts, such as Putnam (1993), that take precisely the opposite tack, arguing that patterns of social exchange are self-perpetuating over the long term, with nearly ineluctable positive or negative consequences for political and economic development. Under Putnam's account, societies with high initial endowments of social capital (which includes appropriate institutions) are likely to prosper, even several centuries after that endowment was established. Societies with low endowments of social capital, in contrast, are likely to remain mired in corruption, distrust, and misery. Thus, southern Italy has suffered through centuries of squalor in the wake of the Norman invasion, while northern and central Italy have done far better because of the legacy of the Renaissance city-states. While Putnam claims in the conclusions of *Making Democracy Work* that it is possible to increase social capital and free southern Italy from the trap that history has set for it, he does not say how, and his optimism sits very awkwardly with the unabashedly pessimistic claims set out in the main part of the book.

Both of these theories invoke large-scale secular forces of development as the main driving factor behind changes in trust and cooperation. As a result, they have poorly developed microfoundations and respectively make overly deterministic predictions about the ineluctability (*Gemeinschaft/Gesellschaft* accounts) or near impossibility (endowment of social capital arguments) of change in how actors trust each other. The theory that I present provides an alternative explanatory framework that does not appeal to broad claims about social change so much as to quite specific arguments about the microfoundations of both trust and institutions. There is no implication of an ineluctable movement from community-based forms of organization based on informal institutions to more impersonal forms based on formal rules, nor of an initial endowment of social capital that sets patterns of social interaction in concrete over the *longue durée*. Instead, the focus is on how particular institutions may change, depending on the relative bargaining power of actors. Thus, it provides a much more specific account of the circumstances under which trust and cooperation will change than rival explanatory frameworks. If patterns of trust may be shaped by institutions, and institutions may in turn change as actors' relative bargaining power changes, then changes in patterns of trust and cooperation are neither inevitable and irreversible, as in the one account, nor impossible, as in the other. We may expect that as relative bargaining power changes so too will informal rules and the patterns of trust and cooperation that depend upon them.

Furthermore, an account of how institutional change affects trust and cooperation can make an important contribution to our more general understanding of change in the political economy. Much recent work explores the ways in which global forces affect the sustainability of more cooperative forms of capitalism, such as those found in many Western European countries. However, there is very little examination indeed in the literature of the role that informal institutions play in market regulation or of how informal institutions may change (Helmke and Levitsky 2004) as a result of changing global conditions (Culpepper 2005). This is a serious gap in the current literature in comparative political economy, and especially the literature on varieties of capitalism.

Chapter 5 provides an account of how material interests and bargaining strength may explain informal institutional change. It marks a substantial advance on existing work in that it details the specific mechanisms through which informal institutional change, as opposed to formal institutional change, is likely to take place. By showing how shifts in power relations may lead to changes in informal institutions that in turn have knock-on consequences for trust and cooperation, we can see how mechanisms of informal institutional change have real consequences for the political economy. Clearly, there are other mechanisms through which informal institutional change may occur. In particular, the account offered in Chapter 5 emphasizes power relations at the expense of ideational transformations (Culpepper 2005). The latter may be of considerable importance in circumstances where actors are unsure about the respective consequences of different kinds of institutional change (Culpepper 2005). Nonetheless, by emphasizing bargaining power, it provides an important corrective to the emphasis on coordination and generalized benefits that characterizes much of the literature on "varieties of capitalism" and a supplement to those theories that similarly seek to arrive at a better understanding of how the interplay of particular interests may culminate in institutional change (Hall and Thelen 2007).

7.5 CONCLUSIONS

This book has sought to unravel the tangled relationship between institutions, trust, and cooperation and thus to reach a better understanding of when institutions will and will not support the rich kinds of trust that many social scientists are interested in. It has used industrial districts – arguably one of the most important practical examples that we have of trust and cooperation in the political economy – as test cases for this theory, showing how institutions seem to play a highly important role in shaping the ways in which individuals in these districts trust and do not trust each other over important forms of cooperation. This test is not meant to be exhaustive. It does, however, suggest that the arguments I make are likely to have wider applicability, insofar as the cases chosen are "difficult" cases for the theoretical methods I employ. A rational choice approach to the understanding of institutions, trust, and cooperation should provide insights across a wide variety of cases and settings.

Furthermore, it provides a better *comparative* understanding of variation between cases than important rival accounts. Jackman and Miller (2004) have argued that rational choice institutionalism is likely to provide a better basis for comparative political research than the political culture approach. I dissent in part from their claims – it seems clear to me that the study of culture (understood as a set of constraints upon individuals' understanding of what is, or is not, socially possible) provides important insights into actors' perceptions and behavior that are beyond the scope of simple rational choice accounts. However, for the purposes of straightforward comparison (as opposed to holistic efforts to reconstruct the *Verstehen* of particular social groups), just such simple accounts have considerable value as a basis for testable assertions regarding the kinds of variation in trust and cooperation that are likely to be found in different institutional settings.

Finally, a proper understanding of the relationship between institutions and trust provides insights into how trust and cooperation are likely to change over time. If institutional change is likely to be associated with changes in how individuals trust or do not trust each other, we can reach a better understanding of key questions of economic and political development that extend far beyond the specific cases under examination here.

Advancing this research agenda into these new areas is beyond the scope of my task here. Still, at least, I hope that this book has provided at least the beginnings of a better understanding of the relationship between institutions and trust. As I have sought to emphasize throughout this book, we are unlikely to understand this relationship unless we clarify the concepts that we use in our attempts to capture it. The main body of this book has sought to do just that, and to demonstrate the applicability of these concepts both to important cases of cooperation in the political economy, and to key theoretical debates in the social sciences. While it does not claim to have resolved all of the ambiguities in the relationship between institutions, trust, and cooperation, it does represent a genuine advance in our understanding of this complex interplay.

Bibliography

Alacevich, Franco, Filippo Bucarelli, Simone Capelli, and Andrea Valzania. 2000. *Emergenza Occupazione: Ruolo delle Parti Sociali, Concertazione e Contrattazione Collettiva in Italia*. Florence: Edizioni CUSL.

Albert, Michel. 1993. *Capitalism against Capitalism*. London: Whurr.

Almond, Gabriel. 1956. "Comparative Political Systems." *Journal of Politics* 18 (4): 391–409.

Almond, Gabriel A., and Sidney Verba. 1963. *The Civic Culture: Political Attitudes and Democracy in Five Nations*. Princeton, N.J.: Princeton University Press.

Amable, Bruno. 2003. *The Diversity of Modern Capitalism*. Oxford: Oxford University Press.

Amin, Ash. 1993. "The Globalization of the Economy: An Erosion of Regional Networks?" In *The Embedded Firm: On the Socioeconomics of Industrial Networks*. Edited by G. Grabher, 278–295. London: Routledge.

1994. "The Difficult Transition from Informal Economy to Marshallian Industrial District." *Area* 26 (1): 13–24.

1998. *Il Modello Emiliano: Sfide Istituzionali*. Bologna: Istituto per il Lavoro.

Amin, Ash, and Nigel Thrift. 1992. "Neo-Marshallian Nodes in Global Networks." *International Journal of Urban and Regional Research* 16 (4): 571–587.

1994. "Living in the Global." In *Globalization, Institutions and Regional Development in Europe*. Edited by A. Amin and N. Thrift 1–22. Oxford: Oxford University Press.

Arlacchi, P. 1988. *Mafia Business: The Mafia Ethic and the Spirit of Capitalism*. Oxford: Oxford University Press.

1993. *Men of Dishonor: Inside the Sicilian Mafia*. New York: William Morrow.

Arrighetti, Alessandro, Reinhard Bachmann, and Simon Deakin. 1997. "Contract Law, Social Norms and Inter-firm Cooperation." *Cambridge Journal of Economics* 21: 171–195.

Arthur, Brian. 1994. *Increasing Returns and Path Dependence in the Economy*. Ann Arbor: University of Michigan Press.

Bagnasco, Arnaldo. 1977. *Tre Italie: La Problematica Territoriale dello Sviluppo Italiano*. Bologna: Il Mulino.

1978. "Problematiche dello Sviluppo e Articolazione dell'Analisi: Un Paradigma per l'Analisi Territoriale." In *Le Problematiche dello Sviluppo Italiano*. Edited by A. Bagnasco, M. Messori and C. Trigilia 205–251. Milan: Feltrinelli.

1985. "La Costruzione Sociale del Mercato: Strategie di Impresa e Esperimenti di Scala in Italia." *Stato e Mercato* 13: 9–45.

1994. "Regioni, Tradizione Civica, Modernizzazione Italiana: Un Commento alla Ricerca di Putnam." *Stato e Mercato* 40: 93–103.

1995. "Introduction: An Unexpected and Controversial Return." In *Small and Medium-Size Enterprises*. Edited by A. Bagnasco and C.F. Sabel 1–14. London: Pinter.

1999. "Teoria del Capitale Sociale e Political Economy Comparata." *Stato e Mercato* 3: 351–372.

Bagnasco, Arnaldo, and Marcello Messori. 1978. "Problematiche dello Sviluppo e Questione della Piccola Impresa." In *Le Problematiche dello Sviluppo Italiano*. Edited by A. Bagnasco, M. Messori and C. Trigilia 7–63. Milan: Feltrinelli.

Bagnasco, Arnaldo, and Carlo Trigilia. 1993. *La Construction Sociale du Marché: Le Défi de la Troisième Italie*. Cachan: Editions de l'ENS-cachan.

Banfield, Edward. 1958. *The Moral Basis of a Backward Society*. Chicago: Free Press.

Barber, B. R. 1996. *Jihad vs. McWorld: How Globalism and Tribalism Are Re-Shaping the World*. New York: Ballantine.

Barnes, Samuel H. 1988. *Politics and Culture*. Ann Arbor: Center for Political Studies, Institute for Social Research, University of Michigan.

Barry, Brian. 1978. *Sociologists, Economists and Democracy*. Chicago: University of Chicago Press.

Bates, Robert H., Avner Greif, Margaret Levi, Jean-Laurent Rosenthal, and Barry R. Weingast. 1998a. "Conclusion." In *Analytic Narratives*. Edited by R.H. Bates, A. Greif, M. Levi, J.-L. Rosenthal, and B.R. Weingast 231–38. Princeton, N.J.: Princeton University Press.

1998b. "Introduction." In *Analytic Narratives*. Edited by R.H. Bates, A. Greif, M. Levi, J.-L. Rosenthal, and B.R. Weingast 3–22. Princeton, N.J.: Princeton University Press.

Batz, Ulrike, Ralf Grammel, Frank Iwer, Vogl, Gerline, and Jürgen Volkert. 1998. *Strukturbericht 1997/98: Zur Wirtschaftlichen und Beschäftigungspolitischen Lage in der Region Stuttgart*. Stuttgart: Institut für Angewandte Wirtschaftsforschung.

Becattini, Giacomo. 1994. "The Development of Light Industry in Tuscany: An Interpretation." In *Regional Development in a Modern European Economy: The Case of Tuscany*. Edited by R. Leonardi and R.Y. Nanetti, 77–94. London: Pinter.

Becker, Laurence C. 1996. "Trust as Noncognitive Security about Motives." *Ethics* 107: 43–61.

Bednar, J., and S. Page. 2007. "Can Game (s) Theory Explain Culture?: The Emergence of Cultural Behavior Within Multiple Games." *Rationality and Society* 19 (1): 65.

Bellandi, Marco. 1992. "The Incentives to Decentralized Industrial Creativity in Local Systems of Small Firms." *Revue d'Économie Industrielle* 59: 99–110.

1996. "Research Briefing: Innovation and Change in the Marshallian Industrial District." *European Planning Studies* 4 (3): 357–368.

Bellandi, Marco, and Marco Romagnoli. 1994. "Case Study II: Prato and the Textile Industry." In *Regional Development in a Modern European Economy: The Case of Tuscany*. Edited by R. Leonardi and R. Y. Nanetti, 145–169. London: Pinter.

Bellini, Nicola, Maria Grazia Giordani, and Francesca Pasquini. 1990. "The Industrial Policy of Emilia-Romagna: The Business Service Centers." In *The Regions and European Integration: The Case of Emilia-Romagna*. Edited by R. Leonardi and R. Nanetti, 171–185. London: Pinter.

Berger, Suzanne. 1996. "Introduction." In *National Diversity and Global Capitalism*. Edited by S. Berger and R. Dore 1–28. Ithaca, N.Y.: Cornell University Press.

2000. "Globalization and Politics." *Annual Review of Political Science* 3: 43–62.

Berger, Suzanne, and Richard M. Locke. 2001. "Il Caso Italiano and Globalization." *Daedalus* 130 (3): 85–104.

Bernstein, Lisa. 1992. "Opting Out of the Legal System: Extralegal Contractual Relations in the Diamond Industry." *Journal of Legal Studies* 21 (1): 115–157.

Bertini, Silvano. 1998. *Artigiani e Piccole Imprese Bolognesi: Casi di un Sistema Imprenditoriale di Successo*. Bologna: Nomisma.

Bianchi, Patrizio. 1993. "Un Contesto Produttivo Non Nasce dal Nulla: È Fatto di Conoscenze, Regole Sociali, Culture, Innovazioni, Finanziamenti." *Scuolaofficina* 93 (2): 34–37.

Bianchi, Patrizio, and Giuseppina Gualteri. 1990. "Emilia-Romagna and Its Industrial Districts: The Evolution of a Model." In *The Regions and European Integration: The Case of Emilia-Romagna*. Edited by R. Leonardi and R. Nanetti 83–108. London: Pinter.

Boari, Cristina. 2001. *Industrial Clusters, Focal Firms and Economic Dynamism: A Perspective from Italy*. Washington, D.C.: World Bank Institute.

Braczyk, Hans-Joachim. 1996. "Erneuerung einer Industrieregion?" In *Kurswechsel in der Industrie: Lean Production in Baden-Württemberg*. Edited by H. J. Braczyk and G. Schienstock, 11–23. Stuttgart: W. Kohlhammer.

Braczyk, Hans Joachim, and Gerd Schienstock. 1996. *Kurswechsel in der Industrie: Lean Production in Baden-Württemberg.* Stuttgart: W. Kohlhammer.

Braczyk, Hans-Joachim, Gerd Schienstock, and Bernd Steffensen. 1996. "Die Regionalökonomie Baden-Württembergs – Ursachen und Grenzen des Erfolgs." In *Kurswechsel in der Industrie: Lean Production in Baden-Württemberg.* Edited by H.J. Braczyk and G. Schienstock 24–51. Stuttgart: W. Kohlhammer.

Brehm, John, and Scott Gates. 1997. *Working, Shirking and Sabotage.* Ann Arbor: University of Michigan Press.

Brioschi, Francesco, Mario Sole Brioschi, and Giulio Cainelli. 2002. "From the Industrial District to the District Group: An Insight into the Evolution of Capitalism in Italy." *Regional Studies* 36 (9): 1037–1052.

Brusca, Giovanni. 2000. "Deposizione del Collaboratore Giovanni Brusca." Palermo, Italy.

Brusco, Sebastiano. 1982. "The Emilian Model: Productive Decentralisation and Social Integration." *Cambridge Journal of Economics* 6 (2): 167–184.

1990. "La Genesi dell'Idea di Distretto Industriale." In *Distretti Industriali e Cooperazione fra Imprese in Italia.* Edited by F. Pyke, G. Becattini, and W. Sengenberger, 28–33. Florence: Studi e Informazioni.

1992. "Small Firms and the Provision of Real Services." In *Industrial Districts and Local Economic Regeneration.* Edited by F. Pyke and W. Sengenberger, 177–196. Geneva: International Institute for Labour Studies.

1996. "Italy: B – Trust, Social Capital and Local Development: Some Lessons from the Experience of the Italian Districts." In *Networks of Enterprises and Local Development: Competing and Co-operating in Local Productive Systems,* 115–120. Paris: OECD.

Brusco, Sebastiano, and Daniela Bigarelli. 2002. "*Regional Productive Systems in the Knitwear and Clothing Sectors in Italy: Industrial Structure and Training Needs.*" Unpublished paper. Available at http://www.ri.it/doc/ItalianApparelIndustry.pdf (accessed on January 10, 2009).

Brusco, Sebastiano, Giulio Cainelli, Fabrizia Forni, Mauro Franchi, Alberto Malusardi, and Roberto Righetti. 1996. "The Evolution of Industrial Districts in Emilia-Romagna." In *Local and Regional Response to Global Pressure: The Case of Italy and Its Industrial Districts.* Edited by F. Cossentino, F. Pyke and W. Sengenberger, 17–36. Geneva: International Institute for Labour Studies.

Brusco, Sebastiano, and Ezio Righi. 1989. "Local Government, Industrial Policy and Social Consensus: The Case of Modena (Italy)." *Economy and Society* 18 (4): 405–424.

Bryson, John, Peter Wood, and David Keeble. 1993. "Business Networks, Small Firm Flexibility and Regional Development in UK Business Services." *Entrepreneurship and Regional Development* 5 (3): 265–277.

Burroni, Luigi, and Carlo Trigilia. 2001. "Italy: Economic Development through Local Economies." In *Local Economies in Europe: Rise or*

Demise? Edited by C. Crouch, P. Le Galès, C. Trigilia, and H. Voelzkow, 46–78. Oxford: Oxford University Press.

Cainelli, G. 2002. "L'Evoluzione dei Distretti Industriali in Italia." *Quaderni IDSE* 5.

Calvert, Randall L. 1995a. *Explaining Social Order: Internalization, External Enforcement, or Equilibrium?* Working Paper No. 4. Rochester, N.Y.: W. Allen Wallis Institute of Political Economy, University of Rochester.

1995b. "Rational Actors, Equilibrium, and Social Institutions." In *Explaining Social Institutions.* Edited by J. Knight and I. Sened. Ann Arbor: University of Michigan Press.

1995c. "The Rational Choice Theory of Social Institutions: Cooperation, Coordination and Communication." In *Modern Political Economy: Old Topics, New Directions.* Edited by J. S. Banks and E. A. Hanushek, 216–268. Cambridge: Cambridge University Press.

Calvert, Randall L., and James Johnson. 1999. "Interpretation and Coordination in Constitutional Politics." In *Lessons in Democracy.* Edited by E. Hauser and J. Wasilewski 99–138. Rochester, N.Y.: University of Rochester Press.

Campbell, John L. 2004. *Institutional Change and Globalization.* Princeton, NJ: Princeton University Press.

Capecchi, Vittorio. 1997. "In Search of Flexibility: The Bologna Metalworking Industry, 1900–1992." In *World of Possibilities: Flexibility and Mass Production in Western Industrialization.* Edited by C.F. Sabel and J. Zeitlin, 381–418. Cambridge: Cambridge University Press.

Caporaso, J. A. 1997. "Across the Great Divide: Integrating Comparative and International Politics." *International Studies Quarterly* 41 (4): 563–592.

Carey, John M. 2000. "Parchment, Equilibria, and Institutions." *Comparative Political Studies* 33 (6–7): 735–761.

Catanzaro, R. 1985. "Enforcers, Entrepreneurs, and Survivors: How the Mafia Has Adapted to Change." *British Journal of Sociology* 36: 34–57.

Cavalli, Alessandro, and Fabio Luca Cavazza. 2001. "Reflections on Political Culture and the 'Italian National Character.'" *Daedalus* 130 (3): 119–138.

Coase, Ronald. 1991. "The Nature of the Firm (1937)." In *The Nature of the Firm: Origins, Evolution and Development.* Edited by O. E. Williamson and S. G. Winter, 18–33. Oxford: Oxford University Press.

Cook, Karen S., Russell Hardin, and Margaret Levi. 2005. *Cooperation without Trust?* New York: Russell Sage Foundation.

Cooke, Philip, and Kevin Morgan. 1994. "Growth Regions under Duress: Renewal Strategies in Baden-Württemberg and Emilia-Romagna." In *Globalization, Institutions and Regional Development in Europe.* Edited by A. Amin and N. Thrift 91–117. Oxford: Oxford University Press.

1998. *The Associational Economy: Firms, Regions and Innovation.* Oxford: Oxford University Press.

Courault, Bruno A., and Claudine Romani. 1992. "A Reexamination of the Italian Model of Flexible Production from a Comparative Point of View." In *Pathways to Industrialization and Regional Development*. Edited by M. Storper and A. J. Scott, 183–192. London: Routledge.

Crouch, Colin, and Henry Farrell. 2004. "Breaking the Path of Institutional Development: Alternatives to the New Determinism in Political Economy." *Rationality and Society* 16 (1): 5–43.

Crouch, Colin, Patrick Le Galès, Carlo Trigilia, and Helmut Voelzkow. 2001. *Local Economies in Europe: Rise or Demise*. Oxford: Oxford University Press.

2004. *Changing Governance of Local Economies: Responses of European Local Production Systems*. Oxford: Oxford University Press.

Crouch, Colin, and Carlo Trigilia. 1997. "Introduction: The Future of Capitalist Diversity." In *Political Economy of Modern Capitalism*. Edited by C. Crouch and W. Streeck, 1–18. London: Sage.

2001. "Conclusion: Still Local Economies in Global Capitalism?" In *Local Economies in Europe: Rise or Demise?* Edited by C. Crouch, P. Le Galès, C. Trigilia and H. Voelzkow, 212–237. Oxford: Oxford University Press.

Culpepper, Pepper D. 2005. "Institutional Change in Contemporary Capitalism: Coordinated Financial Systems since 1990." *World Politics* 57 (2): 173–199.

Curti, Roberto, and Mauro Grandi. 1997. *Per Niente Fragile: Bologna Capitale del Packaging*. Bologna: Museo del Patrimonio Industriale.

Dasgupta, Partha. 1988. "Trust as a Commodity." In *Trust: Making and Breaking Cooperative Relations*. Edited by D. Gambetta, 49–72. Oxford: Basil Blackwell.

Davies, Daniel. 2006. *Reputations Are Made Of ...* Available at http://crookedtimber.org/2006/11/29/reputations-are-made-of/ (accessed January 10, 2009).

Dei Ottati, Gabi. 1991. "The Economic Bases of Diffuse Industrialization." *International Studies of Management and Organization* 21 (1): 53–74.

1994a. "Case Study I: Prato and Its Evolution in a European Context." In *Regional Development in a Modern European Economy: The Case of Tuscany*. Edited by R. Leonardi and R. Y. Nanetti, 116–144. London: Pinter.

1994b. "Trust, Interlinking Transactions and Credit in the Industrial District." *Cambridge Journal of Economics* 18: 529–546.

1996. "Economic Changes in the District of Prato in the 1980s: Towards a More Conscious and Organized Industrial District." *European Planning Studies* 4 (1): 35–52.

1996. "The Remarkable Resilience of the Industrial Districts of Tuscany." In *Local and Regional Response to Global Pressure: The Case of Italy and Its Industrial Districts*. Edited by F. Cossentino, F. Pyke, and W. Sengenberger, 21–43. Geneva: International Institute for Labour Studies.

2003. "Global Competition and Entrepreneurial Behaviour in Industrial Districts: Trust Relations in an Italian Industrial District." Paper presented at the Conference on Clusters, Industrial Districts and Firms: The Challenge of Globalization, Modena, Italy, September 12–13, 2003.

Dickie, J. 2004. *Cosa Nostra: A History of the Sicilian Mafia*: New York: Palgrave Macmillan.

DiGiovanna, Sean. 1996. "Industrial Districts and Regional Economic Development: A Regulation Approach." *Regional Studies* 30 (4): 373–386.

DiMaggio, Paul. 1998. "The New Institutionalisms: Avenues of Collaboration." *Journal of Institutional and Theoretical Economics* 154 (4): 697–705.

Dixit, Avinash. 2004. *Lawlessness and Economics: Alternative Modes of Governance*. Princeton, N.J.: Princeton University Press.

Dore, Ronald. 1992. "Goodwill and the Spirit of Market Capitalism." In *The Sociology of Economic Life*. Edited by R. Swedberg and M. Granovetter, 425–443. Boulder, Colo.: Westview.

Eckstein, Harry. 1988. "A Culturalist Theory of Political Change." *American Political Science Review* 82: 789–804.

1996. "Culture as a Foundation Concept for the Social Sciences." *Journal of Theoretical Politics* 8 (4): 471–497.

1997. "Social Science as Cultural Science: Rational Choice as Metaphysics." In *Culture Matters: Essays in Honor of Aaron Wildavsky*. Edited by R.J. Ellis and M. Thompson, 21–44. Boulder, Colo.: Westview.

Elkins, David J., and Richard E.B. Simeon. 1979. "A Cause in Search of Its Effect, or What does Political Culture Explain?" *Comparative Politics* 11 (2): 127–146.

Elster, Jon. 1989. *The Cement of Society: A Study of Social Order*. Cambridge: Cambridge University Press.

Ensminger, Jean. 2003. "Reputations, Trust and the Principal-Agent Problem." In *Trust and Society*. Edited by K. Cook, 185–202. New York: Russell Sage.

Ensminger, Jean, and Jack Knight. 1997. "Changing Social Norms: Common Property, Bridewealth, and Clan Exogamy." *Current Anthropology* 38 (1): 1–24.

Espina, Alvaro. 1996. "Local Productive Systems and Structural Change." In *Networks of Enterprises and Local Development: Competing and Co-operating in Local Productive Systems*, 81–89. Paris: OECD.

Falcone, G., and M. Padovani. 1992. *Cose di Cosa Nostra*. Milan: Rizzoli.

Farrell, Henry. 2004. "Trust, Distrust and Power." In *Distrust*. Edited by R. Hardin, 85–105. New York: Russell Sage Foundation.

2005. "Trust and Political Economy: Comparing the Effects of Institutions on Inter-Firm Cooperation." *Comparative Political Studies* 38 (5): 459–483.

Farrell, Henry, and Daniel W. Drezner. 2008. "The Power and Politics of Blogs." *Public Choice* 134: 15–30.

Farrell, Henry, and Adrienne Héritier. 2003. "Formal and Informal Institutions Under Codecision: Continuous Constitution-Building in Europe."

Governance: An International Journal of Policy, Administration, and Institutions 16 (4): 577–600.

Farrell, Henry, and Ann-Louise Holten. 2004. "Collective Goods in the Local Economy: The Packaging Machinery Cluster in Bologna." In *Changing Governance of Local Economies: Responses of European Local Production Systems*. Edited by C. Crouch, P. Le Galès, C. Trigilia, and H. Voelzkow, 23–45. Oxford: Oxford University Press.

Farrell, Henry, and Jack Knight. *Trust and Institutional Compliance*. Unpublished document.

2003. "Trust, Institutions and Institutional Change: Industrial Districts and the Social Capital Hypothesis." *Politics and Society* 31 (4): 537–556.

Feagin, Joe R. 1991. "The Continuing Significance of Race: Antiblack Discrimination in Public Places." *American Sociological Review* 56 (1): 101–116.

Ferrucci, L., and R. Varaldo. 1993. "La Natura e la Dinamica dell'Impresa Distrettuale." *Economia e Politica Industriale* 80: 73–97.

Fligstein, Neil. 1996. "Markets as Politics: A Political-Cultural Approach to Market Institutions." *American Sociological Review* 61: 656–673.

Franchi, Maura, and Vittorio Rieser. 1991. "Le Categorie Sociologiche nell'Analisi del Distretto Industriale: Tra Comunità e Razionalizzazione." *Stato e Mercato* 33: 451–476.

Frieden, J., and L. L. Martin. 2002. "International Political Economy: Global and Domestic Interactions." *Political Science: The State of the Discipline Centennial Edition*: 118–146.

Frieden, J. A. 1991. "Invested Interests: The Politics of National Economic Policies in a World of Global Finance." *International Organization* 45 (4): 425–451.

Friedman, T. 2000. *The Lexus and the Olive Tree*. New York: Basic Books.

Fuchs, Gerhard, and Sandra Wassermann. 2004. *The Regional Innovation System in Baden-Württemberg: Lock-In or Breakthrough?* Stuttgart: University of Stuttgart.

Gambetta, Diego. 1988. "Can We Trust Trust?" In *Trust: Making and Breaking Cooperative Relations*. Edited by D. Gambetta. Oxford: Basil Blackwell.

1993. *The Sicilian Mafia: The Business of Private Protection*. Cambridge, MA: Harvard University Press.

Gamm, Gerald, and Kenneth Shepsle. 1989. "Emergence of Legislative Institutions: Standing Committees in the House and Senate." *Legislative Studies Quarterly* 14 (1): 39–66.

Garrett, Geoffrey, and Peter Lange. 1995. "Internationalization, Institutions, and Political Change." *International Organization* 49 (4): 627–655.

Gibbons, Robert, and Andrew Rutten. 2004. "*Institutional Interactions: An Equilibrium Approach to the State and Civil Society*." Unpublished paper.

Glassmann, Ulrich. 2004. "Refining National Policy: The Machine-Tool Industry in the Local Economy of Stuttgart." In *Changing Governance of Local Economies: Responses of European Local Production Systems*. Edited by C. Crouch, P. Le Galès, C. Trigilia, and H. Voelzkow, 46–73. Oxford: Oxford University Press.

Glassmann, Ulrich, and Helmut Voelzkow. 2001. "The Governance of Local Economies in Germany." In *Local Economies in Europe: Rise or Demise?* Edited by C. Crouch, P. Le Galès, C. Trigilia, and H. Voelzkow, 79–116. Oxford: Oxford University Press.

Gould, Roger V. 2003. "Uses of Network Tools in Comparative Historical Research." In *Comparative Historical Analysis in the Social Sciences*. Edited by James Mahoney and Dietrich Rueschemeyer, 241–269. Cambridge: Cambridge University Press.

Gramsci, Antonio. 1971. *Selections from the Prison Notebooks*. London: Lawrence & Wishart.

Granovetter, Mark. 1973. "The Strength of Weak Ties." *American Journal of Sociology* 78 (6): 1360.

1985. "Economic Action and Social Structure: The Problem of Embeddedness." *American Journal of Sociology* 91: 481–510.

Greider, W. 1997. *One World, Ready or Not: The Manic Logic of Global Capitalism*. New York: Simon and Schuster.

Greif, Avner. 1994. "Cultural Beliefs and the Organization of Society: A Historical and Theoretical Reflection on Collectivist and Individualist Societies." *Journal of Political Economy* 102 (5): 912–950.

1997. "*Self-enforcing Political Systems and Economic Growth: Late Medieval Genoa*." Stanford, Calif.: Stanford University, Department of Economics Working Paper.

1999. "Impersonal Exchange and the Origins of Markets. From the Community Responsibility System to Individual Legal Responsibility in Pre-Modern Europe." In *Communities and Markets*. Edited by M. Aoki and Y. Hayami, 3–41. Oxford: Oxford University Press.

2005. *Institutions and the Path to the Modern Economy: Lessons from Medieval Trade*. Cambridge: Cambridge University Press.

Greif, Avner, and David Laitin. 2004. "A Theory of Endogenous Institutional Change." *American Political Science Review* 98 (4): 633–652.

Greif, Avner, Paul Milgrom, and Barry R. Weingast. 1995. "Coordination, Commitment and Enforcement: The Case of the Merchant Guild." In *Explaining Social Institutions*. Edited by J. Knight and I. Sened, 27–56. Ann Arbor: University of Michigan Press.

Grotz, Reinhold, and Boris Braun. 1997. "Territorial or Trans-territorial Networking: Spatial Aspects of Technology-oriented Co-operation within the German Mechanical Engineering System." *Regional Studies* 31 (6): 545–557.

Hacker, Jacob S. 2005. "Policy Drift: The Hidden Politics of US Welfare State Retrenchment." In *Beyond Continuity: Institutional Change in*

Advanced Political Economies. Edited by W. Streeck and K. Thelen, 40–82. New York: Oxford University Press.

Hall, Peter A. 1999. "The Political Economy of Europe in an Era of Interdependence." In *Continuity and Change in Contemporary Capitalism*. Edited by H. Kitschelt, P. Lang, G. Marks, and J. D. Stephens, 101–163. Oxford: Oxford University Press.

2003. "Aligning Ontology and Methodology in Comparative Research." In *Comparative Historical Analysis*. Edited by J. A. Mahoney and D. Rueschemeyer, 333–372. Cambridge: Cambridge University Press.

2005. "Institutional Complementarities: Causes and Effects." *Socio-Economic Review* 3: 373–377.

Hall, Peter A., and David Soskice. 2001. "Introduction." In *Varieties of Capitalism*. Edited by P. Hall and S. David, 1–68. Oxford: Oxford University Press.

Hall, Peter A., and Rosemary C. R. Taylor. 1996. "Political Science and the Three New Institutionalisms." *Political Studies* 44: 936–957.

Hall, Peter A., and Kathleen Thelen. 2007. "*Institutional Change in Varieties of Capitalism*." Unpublished paper.

Hardin, Russell. 1982. *Collective Action*. Baltimore: Johns Hopkins.

1998. "Trust in Government." In *Trust and Governance*. Edited by V. Braithwaite and M. Levi 9–27. New York: Russell Sage Foundation.

2002. *Trust and Trustworthiness*. New York: Russell Sage Foundation.

Harrison, Bennett. 1992. "Industrial Districts: Old Wine in New Bottles?" *Regional Studies* 26 (5): 469–483.

1994. *Lean and Mean: The Changing Landscape of Corporate Power in the Age of Flexibility*. New York: Basic Books.

Hart, H. L. A. 1994. *The Concept of Law*. Oxford: Oxford University Press.

Hay, C. Douglas, ed. 1975. *Albion's Fatal Tree: Crime and Society in 18th-Century England*. London: Allen Lane.

Heidenreich, Martin. 1996. "Beyond Flexible Specialization: The Rearrangement of Regional Production Orders in Emilia-Romagna and Baden-Württemberg." *European Planning Studies* 4 (4): 401–419.

Heidenreich, Martin, and Gerhard Krauss. 1998. "The Baden-Württemberg Production and Innovation Regime: Past Successes and New Challenges." In *Regional Innovation Systems: The Role of Governances in a Globalized World*. Edited by H.-J. Braczyk, P. Cooke, and M. Heidenreich, 186–213. London: University College London Press.

Heinze, Rolf G., and Schmid, Josef. 1994. "Mesokorportische Strategien im Verleich: Industrieller Strukturwandel und Kontingenz Politisher Steurung in Drei Bundesländern. In *Staat und Verbände*. Edited by Wolfgang Streek. Opladen: Westdeutscher Verlag.

Helmke, Gretchen, and Steven Levitsky. 2004. "Informal Institutions and Comparative Politics: A Research Agenda." *Perspectives on Politics* 2 (4): 725–740.

Helper, Susan. 1993. "An Exit-Voice Analysis of Supplier Relations: The Case of the US Automobile Industry." In *The Embedded Firm: On the Socioeconomics of Industrial Networks*. Edited by G. Grabher, 141–160. London: Routledge.

Herrigel, Gary B. 1990. "*Industrial Organization and the Politics of Industry: Centralized and Decentralized Production in Germany*." Cambridge, Mass.: Massachussetts Institute of Technology, Department of Political Science.

1993. "Power and the Redefinition of Industrial Districts: The Case of Baden-Württemberg." In *The Embedded Firm: On the Socioeconomics of Industrial Networks*. Edited by G. Grabher. London: Routledge.

1996a. *Industrial Constructions: The Sources of German Industrial Power*. Cambridge: Cambridge University Press.

1996b. "Crisis in German Decentralized Production: Unexpected Rigidity and the Challenge of an Alternative Form of Flexible Organization in Baden-Württemberg." *European Urban and Regional Studies* 3 (1): 33–52.

1999. "*De-Regionalization, Re-Regionalization and the Transformation of Manufacturing Flexibility: Large Firms and Industrial Districts in Europe*." Unpublished Paper.

2004. "*Space and Governance in New Old Economy Manufacturing Industries*." Bonn: SECONS Working Paper.

2005. "Institutionalists at the Limits of Institutionalism: A Constructivist Critique of Two Volumes by Wolfgang Streeck and Kozo Yamamura." *Socio-Economic Review* 3: 559–567.

Hirst, Paul, and Jonathan Zeitlin. 1992. "Flexible Specialization versus Post-Fordism: Theory, Evidence, and Policy Implications." In *Pathways to Industrialization and Regional Development*. Edited by M. Storper and A.J. Scott, 63–101. London: Routledge.

Hirst, P.Q., and G. Thompson. 1996. *Globalization in Question: The International Economy and the Possibilities of Governance*. Cambridge: Polity Press.

Hollingsworth, J. Rogers, and Robert Boyer. 1997. "Coordination of Economic Actors and Social Systems of Production." In *Contemporary Capitalism: The Embeddedness of Institutions*. Edited by J.R. Hollingsworth and R. Boyer, 1–48. Cambridge: Cambridge University Press.

Hollingsworth, J. Rogers, Philippe C. Schmitter, and Wolfgang Streeck. 1994. "Capitalism, Sectors, Institutions and Performance." In *Governing Capitalist Economies: Performance and Control of Economic Sectors*. Edited by J.R. Hollingsworth, P.C. Schmitter, and W. Streeck, 3–16. Oxford: Oxford University Press.

Inglehart, Ronald. 1988. "Aggregate Stability and Individual Level Flux in Mass Belief Systems." *American Political Science Review* 82: 97–116.

1990. *Culture Shift in Advanced Industrial Societies*. Princeton, N.J.: Princeton University Press.

1997. *Modernization and Postmodernization.* Princeton, N.J.: Princeton University Press.

Innocenti, A. 1998. "Gerarchia e Contratti. Il Ruolo dei Rapporti di Subfornitura tra Piccole Imprese nell'Evoluzione dei Distretti Industriali. *L'industria* (2): 391–416.

ISTAT (Italian National Institute of Statistics). Http://www.istat.it.

Jackman, Robert W., and Ross A. Miller. 1996a. "The Poverty of Political Culture." *American Journal of Political Science* 40 (3): 697–716.

1996b. "A Renaissance of Political Culture?" *American Journal of Political Science* 40 (3): 632–659.

2004. *Before Norms: Institutions and Civic Culture.* Ann Arbor: University of Michigan Press.

Jessop, Bob. 1992. "Fordism and Post-Fordism: A Critical Reformulation." In *Pathways to Industrialization and Regional Development.* Edited by A.J. Scott and M. Storper, 42–62. London: Routledge.

Johnson, James. 1997. "Symbol and Strategy in Comparative Politics." *APSA-CP: Newsletter of the APSA Organized Section in Comparative Politics* (Summer): 6–8.

2002. "How Conceptual Problems Migrate: Rational Choice, Interpretation, and the Hazards of Pluralism." *Annual Review of Political Science* 5: 223–248.

2003. "Conceptual Problems as Obstacles to Progress in Political Science: Four Decades of Political Culture Research." *Journal of Theoretical Politics* 15 (1): 87–115.

2005. "What the Politics of Enfranchisement Can Tell Us About How Rational Choice Theorists Study Institutions." In *Preferences and Situations: Points of Intersection Between Historical and Rational Choice Institutionalism.* Edited by I. Katznelson and B.R. Weingast, 279–312. New York: Russell Sage Foundation.

Katzenstein, P. J. 1985. *Small States in World Markets: Industrial Policy in Europe.* Ithaca, N.Y.: Cornell University Press.

Katznelson, Ira, and Barry R. Weingast. 2005. "Intersections Between Historical and Rational Choice Institutionalism." In *Preferences and Situations: Points of Intersection Between Historical and Rational Choice Institutionalism.* Edited by I. Katznelson and B. Weingast, 1–26. New York: Russell Sage.

Keohane, Robert O. 1984. *After Hegemony: Cooperation and Discord in the World Political Economy.* Princeton, N.J.: Princeton University Press.

Keohane, Robert O., and H.V. Milner. 1996. *Internationalization and Domestic Politics.* Cambridge: Cambridge University Press.

Kerst, C. and B. Steffensen. 1995. *Die Krise des Baden-Württemburgischen Maschinenbaus im Spiegel des NIFAS-Panels.* Stuttgart: Akademie Für Technikfolgenabschätzung.

Kitschelt, H. 1999. *Continuity and Change in Contemporary Capitalism.* Cambridge: Cambridge University Press.

Knight, Jack. 1992. *Institutions and Social Conflict*. Cambridge: Cambridge University Press.

1995. "Models, Interpretations and Theories: Constructing Explanations of Institutional Emergence and Change." In *Explaining Social Institutions*. Edited by J. Knight and I. Sened, 95–120. Ann Arbor: University of Michigan Press.

1998. "The Bases of Cooperation: Social Norms and the Rule of Law." *Journal of Institutional and Theoretical Economics* 154 (4): 754–763.

2000. "Social Norms and the Rule of Law: Fostering Trust in a Socially Diverse Society." In *Trust in Society*. Edited by K. Cook, 354–373. New York: Russell Sage.

Kogut, Bruce, Weijian Shan, and Gordon Walker. 1993. "Knowledge in the Network and the Network as Knowledge: The Structuring of New Industries." In *The Embedded Firm: On the Socioeconomics of Industrial Networks*. Edited by G. Grabher, 67–94. London: Routledge.

Kramer, Roderick. M. 1994. "The Sinister Attribution Error: Paranoid Cognition and Collective Distrust in Organizations. *Motivation and Emotion* 18 (2): 199–230.

1999. "Trust in Organizations: Emerging Perspectives, Enduring Questions." *Annual Review of Psychology* 50: 569–598.

2002. "Organizational Paranoia: Origins and Dynamics." In *Research in Organizational Behavior* 23. Edited by R.W. Sutton and B.M.I. Staw, 1–42. Amsterdam: New Elsevier.

Krasner, Stephen D. 1991. "Global Communications and National Power: Life on the Pareto Frontier." *World Politics* 43 (3): 336–366.

Kreps, David M. 1990. "Corporate Culture and Economic Theory." In *Perspectives on Positive Political Economy*. Edited by J.E. Alt and K.A. Shepsle, 90–143. Cambridge: Cambridge University Press.

1990. *Game Theory and Economic Modelling*. Oxford: Clarendon Press.

Krugman, Paul. 1991. *Geography and Trade*. Leuven: Leuven University Press.

LaPalombara, Joseph. 1965. "Italy: Fragmentation, Isolation, and Alienation." In *Political Culture and Political Development*. Edited by L. W. Pye and S. Verba, 282–329. Princeton, N.J.: Princeton University Press.

Lazerson, Mark. 1993. "Factory or Putting-Out? Knitting Networks in Modena." In *The Embedded Firm: On the Socioeconomics of Industrial Networks*. Edited by G. Grabher, 203–226. London: Routledge.

Lazerson, Mark, and Gianni Lorenzoni. 1999. "The Firms That Feed Industrial Districts: A Return to the Italian Source." *Industrial and Corporate Change* 8 (2): 235–266.

Leonardi, Robert. 1990. "Political Developments and Institutional Change in Emilia-Romagna, 1970–1990." In *The Regions and European Integration: The Case of Emilia-Romagna*. Edited by R. Leonardi and R. Nanetti, 13–38. London: Pinter.

Levi, Margaret. 1988. *Of Rule and Revenue*. Berkeley: University of California Press.

1996. "Social and Unsocial Capital: A Review Essay of Robert Putnam's Making Democracy Work." *Politics & Society* 24 (1): 45–55.

1997. *Consent, Dissent and Patriotism*. Cambridge: Cambridge University Press.

1998. "A State of Trust." In *Trust and Governance*. Edited by V. Braithwaite and M. Levi, 77–101. New York: Russell Sage.

2004. "An Analytic Narrative Approach to Puzzles and Problems." In *Problems and Methods in the Study of Politics*. Edited by I. Shapiro and R.M. Smith, 201–226. Cambridge: Cambridge University Press.

Lijphart, Arend. 1975. The Comparable-Cases Strategy in Comparative Research. *Comparative Political Studies* 8 (2): 158–177.

Locke, Richard M. 1995. *Remaking the Italian Economy*. Ithaca, N.Y.: Cornell University Press.

2001. "*Building Trust*." Unpublished paper.

Lodato, S. 1999. *Ho Ucciso Giovanni Falcone. La Confessione di Giovanni Brusca*. Milan: Mondadori.

Lorenz, Edward H. 1988. "Neither Friends nor Strangers: Informal Networks of Subcontracting in French Industry." In *Trust: Making and Breaking Cooperative Relations*. Edited by D. Gambetta, 194–210. Oxford: Basil Blackwell.

1992a. "Trust and the Flexible Firm: International Comparisons." *Industrial Relations* 31 (3): 455–472.

1992b. "Trust, Community and Cooperation: Toward a Theory of Industrial Districts." In *Pathways to Industrialization and Regional Development*. Edited by M. Storper and A.J. Scott, 195–204. London: Routledge.

1993. "Flexible Production Systems and the Social Construction of Trust." *Politics and Society* 21 (3): 307–324.

1999. "Trust, Contract and Economic Cooperation." *Cambridge Journal of Economics* 23: 301–315.

Lyons, Bruce, and Judith Mehta. 1997. "Contracts, Opportunism and Trust: Self-Interest and Social Orientation." *Cambridge Journal of Economics* 21: 239–257.

Macaulay, Stuart. 1963. "Non-Contractual Relations in Business." *American Sociological Review* 28: 55–67.

MacNeil, Ian. 1978. "Contracts: Adjustment of Long-term Economic Relations under Classical, Neoclassical, and Relational Contract Law." *Northwestern University Law Review* 72 (6): 854–905.

Mahoney, James. 2000. "Path Dependence in Historical Sociology." *Theory and Society* 29 (4): 507–548.

2003. "Strategies of Causal Assessment in Comparative Historical Analysis." In *Comparative Historical Analysis in the Social Sciences*. Edited by James Mahoney and Dietrich Rueschemeyer. New York: Cambridge University Press.

Malusardi, Alberto. 1998. "Contributo." In *Il Mare del Lavoro: Fare e Disfare nella Rete Globale*. Edited by R. Bennati, 47–52. Milan: Costa e Nolan.

March, James G., and Johan P. Olsen. 1989. *Rediscovering Institutions: The Organizational Basis of Politics*. New York: Free Press.

Markovsky, Barry. 1997. "Network Games." *Rationality and Society* 9 (1): 67–90.

McMillan, John, and Christopher Woodruff. 2000. "Private Order under Dysfunctional Public Order." *Michigan Law Review* 98 (8): 2421–2480.

Milgrom, Paul R., Douglass C. North, and Barry R. Weingast. 1990. "The Role of Institutions in the Revival of Trade: The Medieval Law Merchant, Private Judges, and the Champagne Fairs." *Economics and Politics* 2 (1): 1–23.

Miller, Gary J. 1992. *Managerial Dilemmas: The Political Economy of Hierarchy*. Cambridge: Cambridge University Press.

2000. "Rational Choice and Dysfunctional Institutions." *Governance* 13 (4): 535–547.

2001. "Why is Trust Necessary in Organizations? The Moral Hazard of Profit Maximization." In *Trust in Society*. Edited by K. Cook. New York: Russell Sage.

2004. "Monitoring, Rules, and the Control Paradox: Can the Good Soldier Svejk be Trusted?" In *Trust and Distrust in Organizations: Dilemmas and Approaches*. Edited by R.M. Kramer and K.S. Cook. New York: Russell Sage Foundation.

Milner, H. V. 1997. *Interests, Institutions, and Information: Domestic Politics and International Relations*. Princeton, N.J.: Princeton University Press.

Moe, Terry. 1987. "Institutions, Interests and Positive Theory: The Politics of the NLRB." *Studies in American Political Development* 2: 236–302.

2005. "Power and Political Institutions." *Perspectives on Politics* 3 (2): 215–233.

Morgan, K. 2004. "The Exaggerated Death of Geography: Learning, Proximity and Territorial Innovation Systems." *Journal of Economic Geography* 4: 3–21.

Müller-Jentsch, Walther, Norbert Malanowski, Beate Seitz, and Ina Drescher. 1998. *Betriebsräte-Befragung und Analyse der Industriepolitik im Maschinen- und Anlagenbau, "Moderniseirung von Arbeitssystemen und Industriellen Beziehungen im Maschinenbau": Abschlussbericht zum DFG-Forschungprojekt (Mu572/5–1)*. Bochum: Ruhr-Universität.

Nannestad, P. 2008. "What Have We Learned About Generalized Trust, If Anything?" *Annual Review of Political Science* 11: 413–436.

Nardin, Giuseppe. 1993. *Picole Imprese e Internazionalizzazione: Il Caso dell'Industria Meccanica Agricola dell'Emilia-Romagna*. Milan: FrancoAngeli.

Newman, Mark E., Albert-Lázló Barabási, and Duncan Watts. 2006. *The Structure and Dynamics of Networks*. Princeton, N.J.: Princeton University Press.

North, Douglass C. 1981. *Structure and Change in Economic History*. New York: Norton.

1990. *Institutions, Institutional Change and Economic Performance*. Cambridge: Cambridge University Press.

North, Douglass C., and Barry R. Weingast. 1995. "Constitutions and Commitment: The Evolution of Institutions Governing Public Choice in Seventeenth-Century England." In *Empirical Studies in Institutional Change*. Edited by L.J. Alston, T. Eggertsson, and D.C. North, 134–165. Cambridge: Cambridge University Press.

Ohmae, K. 1990. *The Borderless World: Power and Strategy in the Interlinked Economy*. New York: Harper Business.

Olson, Mancur. 1968. *The Logic of Collective Action*. New York: Schocken.

Ordeshook, Peter C. 1990. "The Emerging Discipline of Political Economy." In *Perspectives on Positive Political Economy*. Edited by J.E. Alt and K.A. Shepsle, 9–30. Cambridge: Cambridge University Press.

Ostrom, Elinor. 1990. *Governing the Commons: The Evolution of Institutions for Collective Action*. Cambridge: Cambridge University Press.

Padgett, John, and Christopher K. Ansell. 1993. "Robust Action and the Rise of the Medici." *American Journal of Sociology* 98 (6): 1259–1319.

Pagden, Anthony. 1988. "The Destruction of Trust and its Economic Consequences in the Case of Eighteenth Century Naples." In *Trust: Making and Breaking Cooperative Relations*. Edited by D. Gambetta, 127–141. Oxford: Basil Blackwell.

Page, Scott E. 2007. *The Difference: How the Power of Diversity Creates Better Groups, Firms, Schools, and Societies*. Princeton, N.J.: Princeton University Press.

Paoli, Letizia. 2000. *Mafia Brotherhoods: Organized Crime, Italian Style*. Oxford: Oxford University Press.

Pateman, Carol. 1980. "The Civic Culture: A Philosophic Critique." In *The Civic Culture Revisited*. Edited by G. Almond and S. Verba, 57–102. New York: Little Brown.

Perulli, Paolo. 1989. "Il Distretto Industriale di Modena." In *Strategie di Riaggiustamento Industriale*. Edited by M. Regini and C.F. Sabel, 249–282. Bologna: Il Mulino.

Pierson, Paul. 2000. "The Limits of Design: Explaining Institutional Origins and Change." *Governance* 13 (4): 475–499.

2004. *Politics in Time: History, Institutions and Social Analysis*. Princeton, N.J.: Princeton University Press.

Piore, Michael. 1992. "Technological Trajectories and the Classical Revival in Economics." In *Pathways to Industrialization and Regional Development*. Edited by M. Storper and A.J. Scott, 140–151. London: Routledge.

Piore, Michael J., and Charles F. Sabel. 1984. *The Second Industrial Divide: Possibilities for Prosperity*. New York: Basic Books.

Polanyi, Karl. 1992. "The Economy as Instituted Process." In *The Sociology of Economic Life*. Edited by R. Swedberg and M. Granovetter, 31–50. Boulder, Colo.: Westview.

Przeworski, Adam, and Henry Teune. 1970. *The Logic of Comparative Social Inquiry*. New York: Wiley-Interscience.

Putnam, Robert D. 1993. *Making Democracy Work: Civic Traditions in Modern Italy*. Princeton, N.J.: Princeton University Press.

Pye, Lucian W. 1965. "Introduction: Political Culture and Political Development." In *Political Culture and Political Development*. Edited by L.W. Pye and S. Verba, 3–26. Princeton, N.J.: Princeton University Press.

Pyke, Frank, and Werner Sengenberger. 1992. "Industrial Districts and Local Economic Regeneration: Research and Policy Issues." In *Industrial Districts and Local Economic Regeneration*. Edited by F. Pyke and W. Sengenberger, 3–29. Geneva: International Institute for Labour Studies.

Regini, Marino. 1983. "Le Condizioni dello Scambio Politico: Nascita e Declino della Concertazione in Italia e Grand Bretagna." *Stato e Mercato* 9: 353–384.

1995. "La Varietà Italiana di Capitalismo: Istituzioni Sociali e Struttura Produttiva negli Anni Ottanta." *Stato e Mercato* 43: 3–26.

1997. "Social Institutions and Production Structure: The Italian Variety of Capitalism in the 1980s." In *Political Economy of Modern Capitalism*. Edited by C. Crouch and W. Streeck, 102–116. London: Sage.

Regini, Marino, and Ida Regalia. 1997. "Employers, Unions and the State: The Resurgence of Concertation in Italy?" In *Crisis and Transition in Italian Politics*. Edited by M. Bull and M. Rhodes, 210–230. London: Frank Cass.

Riker, William. 1980. "Implications from the Disequilibrium of Majority Rule for the Study of Institutions." *American Political Science Review* 74 (2): 432–446.

Rinaldi, Alberto. 2002. *The Emilian Model: Twenty Years After*. Unpublished paper. Department of Political Economy, University of Modena.

Rogowski, R. 1989. *Commerce and Coalitions: How Trade Affects Domestic Political Alignments*. Princeton, N.J.: Princeton University Press.

Rothstein, Bo, and Eric Uslaner. 2005. "All for All: Equality and Social Trust." *World Politics* 58: 41–72.

Sabel, Charles F. 1993. "Studied Trust: Building New Forms of Cooperation in a Volatile Economy." In *Explorations in Economic Sociology*. Edited by R. Swedberg, 104–144. New York: Russell Sage Foundation.

2004. "Mondo in Bottiglia o Finestra sul Mondo? Domande Aperte sui Distretti Industriali nello Spirito di Sebastiano Brusco. *Stato e Mercato* (70): 143–158.

Sabel, Charles F., Gary B. Herrigel, Richard Deeg, and Richard Kazis. 1989. "Regional Prosperities Compared: Massachusetts and Baden-Württemberg in the 1980s." *Economy and Society* 18 (4): 374–404.

Sabetti, Filippo. 1996. "Path Dependency and Civic Culture: Some Lessons From Italy About Interpreting Social Experiments." *Politics and Society* 24 (1): 19–44.

2000. *The Search for Good Government: Understanding the Paradox of Italian Democracy*. Montreal: McGill-Queen's University Press.

Sahlins, Marshall. 1972. *Stone Age Economics*. New York: De Gruyter.

1976. *Culture and Practical Reason*. Chicago: University of Chicago.

Sako, Mari. 1991. "The Role of 'Trust' in Japanese Buyer-Supplier Relationships." *Richerche Economiche* 45 (2–3): 449–474.

Sako, Mari, and Susan Helper. 1998. "Determinants of Trust in Supplier Relations: Evidence from the Automotive Industry in Japan and the United States." *Journal of Economic Behavior & Organization* 34 (3): 387–417.

Sartori, Giovanni. 1984. "Guidelines for Concept Analysis." In *Social Science Concepts: A Systematic Analysis*. Edited by G. Sartori, 15–85. Beverly Hills, Calif.: Sage.

Saxenian, Anna Lee. 2006. *The New Argonauts: Regional Advantage in a Global Economy*. Cambridge, Mass.: Harvard University Press.

Schelling, Thomas C. 1960. *The Strategy of Conflict*. Cambridge, Mass.: Harvard University Press.

Schiemann, John W. 2000. "Meeting Halfway Between Rochester and Frankfurt: Generative Salience, Focal Points and Strategic Interaction." *American Journal of Political Science* 44 (1): 1–16.

Schmidt, Vivien. 2002. *The Futures of European Capitalism*. Oxford: Oxford University Press.

Schneiberg, M. 2007. "What's on the path? Path dependence, organizational diversity and the problem of institutional change in the US economy, 1900–1950." *Socio-Economic Review* 5 (1): 47.

Schotter, Andrew. 1981. *The Economic Theory of Social Institutions*. Cambridge: Cambridge University Press.

Sciarrone, R. 1998. *Mafie Vecchie, Mafie Nuove: Radicamento Ed Espansione*. Rome: Donzelli Editore.

Scott, Allen J., and Michael Storper. 1992. "Industrialization and Regional Development." In *Pathways to Industrialization and Regional Development*. Edited by A.J. Scott and M. Storper, 3–18. London: Routledge.

2003. "Regions, Globalization, Development." *Regional Studies* 37 (6): 549–578.

Semlinger, Klaus. 1993a. "Economic Development and Industrial Policy in Baden-Württemberg: Small Firms in a Benevolent Environment." *European Planning Studies* 1 (4): 435–463.

1993b. "Small Firms and Outsourcing as Flexibility Reservoirs of Large Firms." In *The Embedded Firm: On the Socioeconomics of Industrial Networks*. Edited by G. Grabher, 161–178. London: Routledge.

1995. "Industrial Policy and Small-Firm Cooperation in Baden-Württemberg." In *Small and Medium-Size Enterprises*. Edited by A. Bagnasco and C.F. Sabel, 15–30. London: Pinter.

Sforzi, Fabio. 1996. "Italy: A – Local Systems of Small and Medium Sized Firms and Industrial Changes." In *Networks of Enterprises and Local Development: Competing and Co-operating in Local Productive Systems*, 99–113. Paris: OECD.

Shepsle, Kenneth A. 1989. "Studying Institutions: Some Lessons from the Rational Choice Approach." *Journal of Theoretical Politics* 1 (2): 131–147.

Shepsle, Kenneth A., and Barry Weingast. 1981. "Structure-Induced Equilibrium and Legislative Choice." *Public Choice* 37 (3): 503–519.

Shmitz, Hubert. 1992. "Industrial Districts: Model and Reality in Baden-Württemberg, Germany." In *Industrial Districts and Local Economic Regeneration*. Edited by F. Pyke and W. Sengenberger, 87–121. Geneva: International Institute for Labour Studies.

Shonfield, Andrew. 1969. *Modern Capitalism: The Changing Balance of Public and Private Power*. Oxford: Oxford University Press.

Silvia, S.J., and W. Schroeder. 2007. "Why Are German Employers Associations Declining?: Arguments and Evidence." *Comparative Political Studies* 40 (12): 1433.

Soskice, D. 1999. "Divergent Production Regimes: Coordinated and Uncoordinated Market Economies in the 1980s and 1990s." In *Continuity and Change in Contemporary Capitalism*. Edited by Herbert Kitschelt, Peter Lange, Gary Marks, and John D. Stephens, 101–134. Cambridge: Cambridge University Press.

Staber, Udo. 1996. "Accounting for Variations in the Performance of Industrial Districts: The Case of Baden-Württemberg." *International Journal of Urban and Regional Research* 20: 299–316.

2001. "The Structure of Networks in Industrial Districts." *International Journal of Urban and Regional Research* 25 (3): 537–552.

Stajano, C., ed. 1992. *Mafia. L'Atto d'Accusa dei Giudici di Palermo*. Roma: Editore Riuniti.

Stille, A. 1995. *Excellent Cadavers: The Mafia and the Death of the First Italian Republic*: New York: Pantheon Books.

Stone Sweet, Alec. 1999. "Judicialization and the Construction of Governance." *Comparative Political Studies* 31: 147–184.

2000. *Governing with Judges: Constitutional Politics in Europe*. Oxford: Oxford University Press.

Storper, Michael. 1995. "The Resurgence of Regional Economies, Ten Years Later: The Region as a Nexus of Untraded Interdependencies." *European Urban and Regional Studies* 2 (3): 191–221.

Streeck, Wolfgang. 1992. *Social Institutions and Economic Performance: Studies of Industrial Relations in Advanced Capitalist Economies*. London: Sage.

1997. "German Capitalism: Does It Exist? Can It Survive?" In *Political Economy of Modern Capitalism: Mapping Convergence and Diversity.* Edited by C. Crouch and W. Streeck, 33–54. London: Sage.

2005. "Introduction: Institutional Change in Advanced Political Economies." In *Beyond Continuity: Institutional Change in Advanced Political Economies.* Edited by W. Streeck and K. Thelen, 1–39. New York: Oxford University Press.

Stuttgart, IHK. 1997. *Statistik 97 die Wirtschaft Baden-Württembergs und der Region Stuttgart.* Stuttgart: IHK.

Sugden, Robert. 1995. "A Theory of Focal Points." *The Economic Journal* 105 (May): 533–550.

Tarrow, Sidney. 1996. "Making Social Science Work Across Space and Time: A Critical Reflection on Robert Putnam's Making Democracy Work." *American Political Science Review* 90 (2): 389–397.

Taylor, Michael. 1976. *Anarchy and Cooperation.* London: John Wiley.

1982. *Community, Anarchy and Liberty.* Cambridge: Cambridge University Press.

1987. *The Possibility of Cooperation.* Cambridge: Cambridge University Press.

1989. "Structure, Culture and Action in the Explanation of Social Change." *Politics and Society* 17 (2): 115–162.

2006. *Rationality and the Ideology of Disconnection.* Cambridge: Cambridge University Press.

Thelen, Kathleen M. 1999. "Historical Institutionalism in Comparative Politics." *Annual Review of Political Science* 2: 369–404.

2002. "How Institutions Evolve: Insights from Comparative-Historical Analysis." In *Comparative Historical Analysis in the Social Sciences.* Edited by J. Mahoney and D. Rueschemeyer, 208–240. New York: Cambridge University Press.

2004. *How Institutions Evolve: The Political Economy of Skills in Germany, Britain, the United States and Japan.* Cambridge: Cambridge University Press.

Thelen, Kathleen M., and I. Kume. 2006. "Coordination as a Political Problem in Coordinated Market Economies." *Governance* 19 (1): 11–42.

Thelen, Kathleen M., and Christa van Wijnbergen. 2003. "The Paradox of Globalization: Labour Markets in Germany and Beyond." *Comparative Political Studies* 36 (8): 859–880.

Tilly, Charles. 1985. "War Making and State Making as Organized Crime." In *Bringing the State Back In.* Edited by P.B. Evans, D. Rueschemeyer, and T. Skocpol, 169–191. Cambridge: Cambridge University Press.

1984. *Big Structures, Large Processes, Huge Comparisons.* New York: Russell Sage.

Tönnies, Ferdinand. 1887. *Gemeinschaft und Gesellschaft.* Leipzig: Fues.

Trigilia, Carlo. 1986. *Grande Partiti e Piccole Imprese: Comunisti e Democristiani nelle Regioni a Economia Diffusa.* Bologna: Il Mulino.

1986. Small-Firm Development and Political Subcultures in Italy. *European Sociological Review* 2 (3): 161–175.

1989. "Il Distretto Industriale di Prato." In *Strategie di Riaggiustamento Industriale.* Edited by M. Regini and C.F. Sabel. Bologna: Il Mulino.

1990. *Sviluppo Senza Autonomia: Effetti Perversi delle Politiche nel Mezzogiorno.* Bologna: Il Mulino.

1990. "Il Lavoro e la Politica nei Distretti Industriali della Terza Italia." In *Distretti Industriali e Cooperazione fra Imprese in Italia.* Edited by F. Pyke, G. Becattini and W. Sengenberger, 173–196. Florence: Studi e Informazioni.

1991. "The Paradox of the Region: Economic Regulation and the Representation of Interests." *Economy and Society* 20 (3): 306–327.

1992. "Italian Industrial Districts: Neither Myth Nor Interlude." In *Industrial Districts and Local Economic Regeneration.* Edited by F. Pyke and W. Sengenberger, 33–47. Geneva: International Institute for Labour Studies.

1996. "Dynamisme Privé et Désordre Public: Note sur le Paradoxe du Développement Italien." *Sociologie du Travail* 38 (1): 6–13.

1998. *Sociologia Economica: Stato, Mercato e Società nel Capitalismo Moderno.* Bologna: Il Mulino.

1999. "Capitale Sociale e Sviluppo Locale." *Stato e Mercato* 3: 419–440.

Tsebelis, George, and Geoffrey Garrett. 2001. "The Institutional Determinants of Intergovernmentalism and Supranationalism in the EU." *International Organization* 55 (2): 357–390.

UCIMA (Unione Costruttori Italiani Macchine Automatiche per il Confezionamento e l'Imballaggio). 1998. *Rapporto 1997: Industria Italiana della Macchine Automatiche per il Confezionamento e l'Imballaggio.* Milan: UCIMA.

1999. *Rapporto 1998: Industria Italiana della Macchine Automatiche per il Confezionamento e l'Imballaggio.* Milan: UCIMA.

2003. *Rapporto 2003: Industria Italiana della Macchine Automatiche per il Confezionamento e l'Imballaggio.* Milan: UCIMA.

Varese, F. 2006. "How Mafias Migrate: The Case of the Ndrangheta in Northern Italy." *Law & Society Review* 40 (2): 411–444.

Voelzkow, Helmut. 1999. "Die Governance Regionaler Ökonomien im internationalen Vergleich: Deutschland und Italien." In *Die Bindungen der Globalisierung Interorganisationsbeziehungen im Regionalen und Globalen Wirtschaftsraum.* Edited by Gerhard Fuchs, Gerhard Krauss, and Hans-Georg Wolf. Marburg: Metropolis.

Watts, D.J., and S.H. Strogatz. 2006. "Collective Dynamics of 'Small-world' Networks." *The Structure and Dynamics of Networks.* Edited by Mark

Newman, Albert-Laszlo Barabasi, and Duncan J. Watts, 301–303. Princeton, N.J.: Princeton University Press.

Weingast, Barry R. 2002. "Rational Choice Institutionalism." In *Political Science: The State of the Discipline*. Edited by I. Katznelson and H.V. Milner, 1–26. New York: Norton.

Weitzer, Ronald, and Steven A. Tuch. 1999. "Race, Class and Perceptions of Discrimination by the Police." *Crime and Delinquency* 45 (4): 494–507.

Wellman, Barry, and S.D. Berkowitz. 1988. "Introduction: Studying Social Structures." In *Social Structure: A Network Approach*. Edited by B. Wellman and S.D. Berkowitz, 1–14. Cambridge: Cambridge University Press.

White, Harrison C. 1988. "Varieties of Markets." In *Social Structure: A Network Approach*. Edited by B. Wellman and S.D. Berkowitz, 226–260. Cambridge: Cambridge University Press.

Whitford, Josh. 2001a. "Decline of a model: Challenge and Response in the Italian Industrial Districts." *Economy and Society* 30 (1): 38–65.

2001b. "*Can Productive Cooperation Survive the Emergence of Market Power in Italian Industrial Districts?*" Paper presented at the American Sociological Association Meetings, Chicago.

Whitford, Josh, and Cuz Potter. 2007. "Regional Economies, Open Networks and the Spatial Fragmentation of Production." *Socioeconomic Review* 5 (3): 497–526.

Williams, Bernard. 1988. "Formal Structures and Social Reality." In *Trust: Making and Breaking Cooperative Relations*. Edited by D. Gambetta, 3–13. Oxford: Basil Blackwell.

Williamson, Oliver E. 1975. *Markets and Hierarchies: Analysis and Antitrust Implications*. New York: Free Press.

1985. *The Economic Institutions of Capitalism: Firms, Markets, Relational Contracting*. New York: Free Press.

1991. "The Logic of Economic Organization." In *The Nature of the Firm: Origins, Evolution and Development*. Edited by O.E. Williamson and S.G. Winter, 90–116. Oxford: Oxford University Press.

1993. "Calculativeness, Trust, and Economic Organization." *Journal of Law and Economics* 36: 453–486.

2000. "The New Institutional Economics: Taking Stock, Looking Ahead." *Journal of Economic Literature* 38 (3): 595–614.

Yamamura, Kozo, and Wolfgang Streeck. 2003. *The End of Diversity? Prospects for German and Japanese Capitalism*. Ithaca, N.Y.: Cornell University Press.

Zeitlin, Jonathan. 1992. "Industrial Districts and Local Economic Regeneration: Overview and Comment." In *Industrial Districts and Local Economic Regeneration*. Edited by F. Pyke and W. Sengenberger, 279–294. Geneva: International Institute for Labour Studies.

Index

Other Books in the Series: *(continued from page iii)*

Nancy Bermeo, ed., *Unemployment in the New Europe*
Carles Boix, *Democracy and Redistribution*
Carles Boix, *Political Parties, Growth, and Equality: Conservative and Social Democratic Economic Strategies in the World Economy*
Catherine Boone, *Merchant Capital and the Roots of State Power in Senegal, 1930–1985*
Catherine Boone, *Political Topographies of the African State: Territorial Authority and Institutional Change*
Michael Bratton and Nicolas van de Walle, *Democratic Experiments in Africa: Regime Transitions in Comparative Perspective*
Michael Bratton, Robert Mattes, and E. Gyimah-Boadi, *Public Opinion, Democracy, and Market Reform in Africa*
Valerie Bunce, *Leaving Socialism and Leaving the State: The End of Yugoslavia, the Soviet Union, and Czechoslovakia*
Daniele Caramani, *The Nationalization of Politics: The Formation of National Electorates and Party Systems in Europe*
Kanchan Chandra, *Why Ethnic Parties Succeed: Patronage and Ethnic Headcounts in India*
José Antonio Cheibub, *Presidentialism, Parliamentarism, and Democracy*
Ruth Berins Collier, *Paths toward Democracy: The Working Class and Elites in Western Europe and South America*
Christian Davenport, *State Repression and the Domestic Democratic Peace*
Donatella della Porta, *Social Movements, Political Violence, and the State*
Alberto Diaz-Cayeros, *Federalism, Fiscal Authority, and Centralization in Latin America*
Thad Dunning, *Crude Democracy: Natural Resource Wealth and Political Regimes*
Gerald Easter, *Reconstructing the State: Personal Networks and Elite Identity*
Margarita Estevez-Abe, *Welfare and Capitalism in Postwar Japan: Party, Bureaucracy, and Business*
M. Steven Fish, *Democracy Derailed in Russia: The Failure of Open Politics*
Robert F. Franzese, *Macroeconomic Policies of Developed Democracies*
Roberto Franzosi, *The Puzzle of Strikes: Class and State Strategies in Postwar Italy*

Geoffrey Garrett, *Partisan Politics in the Global Economy*
Scott Gehlbach, *Representation through Taxation: Revenue, Politics, and Development in Postcommunist States*
Miriam Golden, *Heroic Defeats: The Politics of Job Loss*
Jeff Goodwin, *No Other Way Out: States and Revolutionary Movements*
Merilee Serrill Grindle, *Changing the State*
Anna Grzymala-Busse, *Rebuilding Leviathan: Party Competition and State Exploitation in Post-Communist Democracies*
Anna Grzymala-Busse, *Redeeming the Communist Past: The Regeneration of Communist Parties in East Central Europe*
Frances Hagopian, *Traditional Politics and Regime Change in Brazil*
Henry E. Hale, *The Foundations of Ethnic Politics: Separatism of States and Nations in Eurasia and the World*
Gretchen Helmke, *Courts Under Constraints: Judges, Generals, and Presidents in Argentina*
Yoshiko Herrera, *Imagined Economies: The Sources of Russian Regionalism*
J. Rogers Hollingsworth and Robert Boyer, eds., *Contemporary Capitalism: The Embeddedness of Institutions*
John D. Huber and Charles R. Shipan, *Deliberate Discretion? The Institutional Foundations of Bureaucratic Autonomy*
Ellen Immergut, *Health Politics: Interests and Institutions in Western Europe*
Torben Iversen, *Capitalism, Democracy, and Welfare*
Torben Iversen, *Contested Economic Institutions*
Torben Iversen, Jonas Pontussen, and David Soskice, eds., *Unions, Employers, and Central Banks: Macroeconomic Coordination and Institutional Change in Social Market Economies*
Thomas Janoski and Alexander M. Hicks, eds., *The Comparative Political Economy of the Welfare State*
Joseph Jupille, *Procedural Politics: Issues, Influence, and Institutional Choice in the European Union*
Stathis Kalyvas, *The Logic of Violence in Civil War*
David C. Kang, *Crony Capitalism: Corruption and Capitalism in South Korea and the Philippines*
Junko Kato, *Regressive Taxation and the Welfare State*
Robert O. Keohane and Helen B. Milner, eds., *Internationalization and Domestic Politics*
Herbert Kitschelt, *The Transformation of European Social Democracy*
Herbert Kitschelt, *Peter Lange, Gary Marks, and John D. Stephens, eds., Continuity and Change in Contemporary Capitalism*

Herbert Kitschelt, Zdenka Mansfeldova, Radek Markowski, and Gabor Toka, *Post-Communist Party Systems*

David Knoke, Franz Urban Pappi, Jeffrey Broadbent, and Yutaka Tsujinaka, eds., *Comparing Policy Networks*

Allan Kornberg and Harold D. Clarke, *Citizens and Community: Political Support in a Representative Democracy*

Amie Kreppel, *The European Parliament and the Supranational Party System*

David D. Laitin, *Language Repertoires and State Construction in Africa*

Fabrice E. Lehoucq and Ivan Molina, *Stuffing the Ballot Box: Fraud, Electoral Reform, and Democratization in Costa Rica*

Mark Irving Lichbach and Alan S. Zuckerman, eds., *Comparative Politics: Rationality, Culture, and Structure*

Evan Lieberman, *Race and Regionalism in the Politics of Taxation in Brazil and South Africa*

Julia Lynch, *Age in the Welfare State: The Origins of Social Spending on Pensioners, Workers, and Children*

Pauline Jones Luong, *Institutional Change and Political Continuity in Post-Soviet Central Asia*

Doug McAdam, John McCarthy, and Mayer Zald, eds., *Comparative Perspectives on Social Movements*

Beatriz Magaloni, *Voting for Autocracy: Hegemonic Party Survival and Its Demise in Mexico*

James Mahoney and Dietrich Rueschemeyer, eds., *Historical Analysis and the Social Sciences*

Scott Mainwaring and Matthew Soberg Shugart, eds., *Presidentialism and Democracy in Latin America*

Isabela Mares, *The Politics of Social Risk: Business and Welfare State Development*

Isabela Mares, *Taxation, Wage Bargaining, and Unemployment*

Anthony W. Marx, *Making Race, Making Nations: A Comparison of South Africa, the United States, and Brazil*

Bonnie Meguid, *Competition between Unequals: The Role of Mainstream Parties in Late-Century Africa*

Joel S. Migdal, *State in Society: Studying How States and Societies Constitute One Another*

Joel S. Migdal, Atul Kohli, and Vivienne Shue, eds., *State Power and Social Forces: Domination and Transformation in the Third World*

Scott Morgenstern and Benito Nacif, eds., *Legislative Politics in Latin America*

Layna Mosley, *Global Capital and National Governments*
Wolfgang C. Müller and Kaare Strøm, *Policy, Office, or Votes?*
Maria Victoria Murillo, *Labor Unions, Partisan Coalitions, and Market Reforms in Latin America*
Ton Notermans, *Money, Markets, and the State: Social Democratic Economic Policies since 1918*
Aníbal Pérez-Liñán, *Presidential Impeachment and the New Political Instability in Latin America*
Roger Petersen, *Understanding Ethnic Violence: Fear, Hatred, and Resentment in Twentieth-Century Eastern Europe*
Simona Piattoni, ed., *Clientelism, Interests, and Democratic Representation*
Paul Pierson, *Dismantling the Welfare State? Reagan, Thatcher, and the Politics of Retrenchment*
Marino Regini, *Uncertain Boundaries: The Social and Political Construction of European Economies*
Marc Howard Ross, *Cultural Contestation in Ethnic Conflict*
Lyle Scruggs, *Sustaining Abundance: Environmental Performance in Industrial Democracies*
Jefferey M. Sellers, *Governing from Below: Urban Regions and the Global Economy*
Yossi Shain and Juan Linz, eds., *Interim Governments and Democratic Transitions*
Beverly Silver, *Forces of Labor: Workers' Movements and Globalization since 1870*
Theda Skocpol, *Social Revolutions in the Modern World*
Regina Smyth, *Candidate Strategies and Electoral Competition in the Russian Federation: Democracy Without Foundation*
Richard Snyder, *Politics after Neoliberalism: Reregulation in Mexico*
David Stark and László Bruszt, *Postsocialist Pathways: Transforming Politics and Property in East Central Europe*
Sven Steinmo, Kathleen Thelen, and Frank Longstreth, eds., *Structuring Politics: Historical Institutionalism in Comparative Analysis*
Susan C. Stokes, *Mandates and Democracy: Neoliberalism by Surprise in Latin America*
Susan C. Stokes, ed., *Public Support for Market Reforms in New Democracies*
Duane Swank, *Global Capital, Political Institutions, and Policy Change in Developed Welfare States*

Sidney Tarrow, *Power in Movement: Social Movements and Contentious Politics*
Kathleen Thelen, *How Institutions Evolve: The Political Economy of Skills in Germany, Britain, the United States, and Japan*
Charles Tilly, *Trust and Rule*
Daniel Treisman, *The Architecture of Government: Rethinking Political Decentralization*
Lily Lee Tsai, *Accountability without Democracy: How Solidary Groups Provide Public Goods in Rural China*
Joshua Tucker, *Regional Economic Voting: Russia, Poland, Hungary, Slovakia, and the Czech Republic, 1990–1999*
Ashutosh Varshney, *Democracy, Development, and the Countryside*
Jeremy M. Weinstein, *Inside Rebellion: The Politics of Insurgent Violence*
Stephen I. Wilkinson, *Votes and Violence: Electoral Competition and Ethnic Riots in India*
Jason Wittenberg, *Crucibles of Political Loyalty: Church Institutions and Electoral Continuity in Hungary*
Elisabeth J. Wood, *Forging Democracy from Below: Insurgent Transitions in South Africa and El Salvador*
Elisabeth J. Wood, *Insurgent Collective Action and Civil War in El Salvador*